Skin-Close Computing and Wearable Technology

Skin-Close Computing and Wearable Technology

Andrews Samraj

CRC Press
Taylor & Francis Group
Boca Raton London New York

CRC Press is an imprint of the
Taylor & Francis Group, an **informa** business

First edition published 2022
by CRC Press
6000 Broken Sound Parkway NW, Suite 300, Boca Raton, FL 33487-2742

and by CRC Press

2 Park Square, Milton Park, Abingdon, Oxon, OX14 4RN

© 2022 Taylor & Francis Group, LLC

CRC Press is an imprint of Taylor & Francis Group, LLC

Library of Congress Cataloging-in-Publication Data
Names: Samraj, Andrews, author.
Title: Skin-close computing and wearable technology : a key to enthusiastic researchers with human-centric approach / Andrews Samraj.
Description: First edition. | Boca Raton, FL : CRC Press, 2022. | Includes bibliographical references and index.
Identifiers: LCCN 2021026438 | ISBN 9780367512163 (hbk) | ISBN 9780367552565 (pbk) | ISBN 9781003052906 (ebk)
Subjects: LCSH: Wearable computers. | Wearable technology. | Self-help devices for people with disabilities.
Classification: LCC QA76.592 .S26 2022 | DDC 004.167--dc23
LC record available at https://lccn.loc.gov/2021026438

ISBN: 978-0-367-51216-3 (hbk)
ISBN: 978-0-367-55256-5 (pbk)
ISBN: 978-1-003-05290-6 (ebk)

DOI: 10.1201/9781003052906

Typeset in Palatino
by SPi Technologies India Pvt Ltd (Straive)

This book is dedicated to the Omnipotent, Omniscient, and Omnipresent God Almighty for everything He provided me.
I also want to dedicate this book to my parents (Late) Mr. N. Samraj and Mrs. A. Juliet Samraj for the support they provided to me for writing this book. I further dedicate this to my wife H. Ranjini Florence and daughters Dr. A. Blessy Grace and Ms. A. Brasen Grizelda for their continuous encouragement and support, which makes me to enjoy my work.

Contents

Foreword xiii
Preface xv
Acknowledgments: Special Thanks xvii
Author's Biography xix

1. **Basic Wearable Computing Requirements and Advantages** 1
   1.1   Wearable Computing and Wearables 2
   1.2   Wearable Types 4
   1.3   Wearable Computer Examples 5

2. **Ergonomics and Its Benefits** 7
   2.1   Definition 7
   2.2   Why Ergonomics 7
   2.3   Work-Related Injuries and Causes of Sickness 8
   2.4   Ergonomic Importance of the Present 9
   2.5   Ergonomics in Wearables 10
   2.6   Ergonomics and Biometrics 11

3. **Applications of Wearable Technology** 13
   3.1   Wearable Types and Applications 13
         3.1.1   Fashion Wearable (Ornaments, Tattoos) 13
         3.1.2   Sports Gear (Shoes, Knee Cap, etc.) 14
         3.1.3   Comfort Wearables (Pads, Elastic Bands,
                 Thermal Comfort Wearables) 14
         3.1.4   Wearable for Childcare (Baby Carriers) 14
         3.1.5   Wearables for Recreation (Tokens, Dollars,
                 Wristbands, and Tattoos) 15
         3.1.6   Wearable for Road Safety (Helmet, Seat Belts) 15
         3.1.7   Wearable for Life Safety (Life Jackets, Anti-
                 Bomb Suit, Biomedical Suits) 15
         3.1.8   Operational Wearable (Gun Holders, Tool Holders,
                 Armor Belts) 15
         3.1.9   Camouflaging Wearable (Nets, Leafy, Grassy Suits,
                 and Helmets for Snippers) 15
         3.1.10 Wearable for Work Place/Manufacturing/Plantations/
                 Farming/Cooking (Apron, Gloves, Cap) 17
         3.1.11 Wearable to Carry Weapon (Rocket Launcher Belt,
                 Grenade Hooks, Bombs, Suicide Bombs) 17
         3.1.12 Wearable for Protection Against Climate
                 (Puffer Jacket, Warmers, Sweaters) 17

3.1.13   Wearable for Protection Against Water (Swim Suits,
          Dive Suits, Pressure Resistive Suits) 17
3.1.14   Wearable for Protection Against High Pressure/Low
          Pressure (Suits that are Designed for Divers and
          Astronauts) 18
3.1.15   Wearable for Fire Safety (Fire-Resistant Suits)..................18
3.1.16   Wearable for Protection Against Vacuum/
          Cosmic Rays/Heat (Astronaut Suits) 18
3.1.17   Wearable for Protection Against Chemical
          (Anticorrosive Suits)...19
3.1.18   Wearable for Protection Against Germs (Masks to
          Protect From Virus like COVID 19, or VD, STD) 20
3.1.19   Wearable for Protection Against Animals/Fishes
          (Wire Mesh Swim Suits)..21
3.1.20   Wearable for Advertisements (Fancy Add-Ons) 21
3.1.21   Wearables for Magic (Magnetic, Covering,
          Especially Stitched Garments) 21
3.1.22   Wearable for Communication (Gloves Like 5DT)............21
3.1.23   Wearables for Wellness (Stress Busters)..............................22
3.1.24   Wearable for Communication with Vehicles
          (Gloves and Microphones) 22
3.1.25   Wearable for Managing Vision Impairments
          (Spectacles, Lenses, AR Smart Glasses) 22
3.1.26   Wearable for Communication with Devices/Life
          Support (Gesture, Movement Capturing Gadgets) 22
3.1.27   Wearable for Communication with Robots/Cyborgs
          (Gloves, Rings, with Embedded Electrodes, Lights,
          Sensors)..23
3.1.28   Wearable in Warfare (Bulletproof Suits, Metal-Plated
          Suits, Proximity Alarms, Weapon Sensors)........................23
3.1.29   Wearable for Pets 23
3.2   Development of Wearables..23

4.   Simple User Interface Design for Wearable Computing and
     Wearable Technology 25
     4.1   Construction of a Hand Glove with Flex Sensors 25
     4.2   Motion Detection Sensors 26
     4.3   Energy Saver/Rain Protection (Multiutility) 27
     4.4   Wearable Band 30

5.   Materials and Components (Constructing a Simple Wearable
     Technology Device) 31
     5.1   Flex/Bend Sensors 31
     5.2   Accelerometer/Magnetometer/Gyroscope 33

5.3	Vibration Sensors	34
5.4	Pressure Sensors/Force Sensors	36
5.5	Smoke/Fire Sensors	36
5.6	Moisture Sensors	37
5.7	Design Considerations	37

6. Assistive Technology and WT 39
6.1 Wearable Electrodes for BCI/BMI 39
    6.1.1 EEG Rhythms 43
    6.1.2 Signal Acquisition by Wearable Methods 43
6.2 BCI – Example 1: Single Trial Source Separations of Visual
    Evoked Potential Signals Using Various Methods and
    Techniques to Identify Controls from Alcoholics 44
    6.2.1 The Methods Involved in This BCI Demonstration 48
    6.2.2 Artificial VEP Simulation 48
        6.2.2.1 Applying Principal Component Analysis 49
    6.2.3 Signal-to-Noise Ratio Calculation 50
    6.2.4 Single Trial P3 Responses Experiment Using
        Real VEP 50
    6.2.5 Outcome of the BCI Experiment 1 52
6.3 BCI Example 2: Speller Paradigm 58
6.4 BCI Example 3: Picture Paradigm 59
6.5 Example 4: Bio-Cyber Machine Gun – A New Mode of
    Authentication Access Using Visual Evoked Potentials 59
6.6 BCI Example 5a: Motor Imagery – Adaptive Bandpass Filter 62
    6.6.1 Demonstration Method 65
    6.6.2 Description of the Dataset 65
    6.6.3 Preprocessing and Feature Extraction 65
    6.6.4 Classification of the Features 67
    6.6.5 Communication with OPC 68
    6.6.6 Experimental Results and Discussion 69
    6.6.7 Consolidation of BMI Design 69
    6.6.8 BCI Example 6b: Motor Imagery – Fractal
        Dimensions in Estimating Features 72
        6.6.8.1 Data and Method of Operation 74
        6.6.8.2 Feature Extraction by Assorted
            Methods of FD 74
        6.6.8.3 TDFD Method 75
        6.6.8.4 DFD and DS Methods 75
        6.6.8.5 Classification 76
        6.6.8.6 Findings 76
6.7 Sign Language 78
    6.7.1 Need for Sign Language 78
    6.7.2 Sign Language Automation 79

          6.7.3   Classes of Sign Language Automation 81
          6.7.4   The Hand Glove or Data Glove 81
          6.7.5   Advantages in Glove-Based Sign Language System 82
    6.8   Wearables in Treatment 85
    6.9   The Wearable Assistive Device PhysiofastHeal..........................86
          6.9.1   The Need and Challenge for the Invention
                  of the Device 87
          6.9.2   Beneficiaries 87
          6.9.3   System Design 87
          6.9.4   Technology Design..89
          6.9.5   The Innovation Introduced in PhysiofastHeal 90
          6.9.6   The Uniqueness of Innovation in Comparison to
                  the Others in This Sector..91
    6.10 Wearables for Blind Population 93

7. **Security Technology and WT**...95
    7.1   Signature Verification 96
          7.1.1   Introduction to Online Signature Verification 96
    7.2   Augmented/Robust Signature Verification 99
          7.2.1   The Modifications in Equipment Setup..........................100
          7.2.2   Subjects and Signal Acquisition Methods 101
          7.2.3   Experimental Setup 102
          7.2.4   Preprocessing and Feature Vector Construction 103
          7.2.5   Matching and Classification..104
          7.2.6   Results and Discussions..104
          7.2.7   Conclusions..106
    7.3   Emergency Response System for Elderly/Disabled/Persons
          in Ambulance...106
          7.3.1   Data Description 107
          7.3.2   Wearable Emergency Response System 108
          7.3.3   Feature Extraction Techniques for System
                  Implementation...109
                  7.3.3.1   Singular Value Decomposition 109
                  7.3.3.2   Fractal Dimension (FD)......................................109
                  7.3.3.3   KATZ'S Method..110
                  7.3.3.4   Fast Fourier Transformations (FFT) 110
                  7.3.3.5   SVD with Average Distribution 110
          7.3.4   Classification Techniques..111
                  7.3.4.1   Euclidean Distance 111
                  7.3.4.2   Linear Discriminant Analysis............................111
          7.3.5   Association of Paradigms 111
                  7.3.5.1   Association Between Paradigms.........................111
                  7.3.5.2   Results of the Experiments 112
                  7.3.5.3   Discussions 112

7.3.5.4 Recommendations Based on the Experiment... 113
7.3.6 Intratrial Variability 113
    7.3.6.1 Discussions and Recommendations
        Based on ITV.. 113
    7.3.6.2 Optimization of This Work................................ 114
    7.3.6.3 Singular Value Decomposition with
        Random Average Distribution............................ 115
    7.3.6.4 Contribution of the Research Work.................... 116

8. Strategic Operation Technology and WT 117
8.1 The Need for Simple and Fast Feature Identification.................. 117
    8.1.1 Modeling the Signal by Zone.............................. 118
    8.1.2 Standard Deviation as a Comparison Feature.................. 119
    8.1.3 Results of Comparison Features 119
    8.1.4 Promises Found in the Wearable Strategic
        Operation Technology... 121
8.2 Consequences of Strategic Operational Technology Using
    Wearable Technology.. 122
8.3 Fast and Easy Gesture Recognition by Wearable in
    Strategic Operations 123
    8.3.1 Mode of Experiments 124
    8.3.2 Use of Singular Value Decomposition (SVD)
        to Quantify Feature Set 127
    8.3.3 Use of Fractal Dimension (FD) to Quantify
        the Feature Set 127
    8.3.4 Results of the Experiment.................................. 128
    8.3.5 Discussion on the Experiment and Its Results 133
    8.3.6 Consolidation of Gesture Movement....................... 138
    8.3.7 Recommendations 141
    8.3.8 Discussions 141
    8.3.9 Conclusion of the Experiment............................. 142
8.4 Zero Error Wearable Technology Applications............................ 142

9. Software and Power Requirements of WT 143
9.1 Writing Codes for Wearable Devices....................................... 143
9.2 Platforms for Development 144
9.3 Understanding the Components of a Device in Operation 145
9.4 Power Requirements for Wearables 146
9.5 Energy Harvesting in Wearables for Self-power 146

10. Higher-Order Human–Robot Interface 149
10.1 Human Interface and Robotic Interface.................................... 149
10.2 Magic Ring for Recognition.. 150
10.3 Authentication of Sex Robot Access Using Wearables 150

10.3.1   The Problem of the Future Generation 151
10.3.2   The Reason for Focus Toward Robotic Emotions............153
10.3.3   Related Examples..153
10.3.4   The Era of Personal Robots and Their Characters 154
         10.3.4.1   Need for Personal Robots 154
         10.3.4.2   Gender Approaches – Transgender Robots.....155
10.3.5   Is It Possible to Have Affective Approaches
         with Robots? 156
         10.3.5.1   Creation of Feelings...156
         10.3.5.2   Appreciation and Rewards: How Robots
                     View This?...157
         10.3.5.3   Psychological and Sociological Approaches...157
         10.3.5.4   Rights of Robots? and Robo-ethics....................158
10.3.6   Areas Where Humans Excel the Robots............................158
         10.3.6.1   Preach – Cannot Repent or Realize 158
         10.3.6.2   Indulge Carnal Pleasure – Cannot
                     Produce Own Children 158
         10.3.6.3   Remember – Cannot Imagine..............................159
         10.3.6.4   Say a Lot of Stories – Cannot Write
                     One by Own...159
         10.3.6.5   Sense and Show Expressions – Cannot Feel
                     Anything, Empathy Sympathy Joy, etc...........159
         10.3.6.6   Restricted to Only the Language of
                     Operation – Humans Can Learn an
                     Unknown Language with Time 160
         10.3.6.7   Can Control Animals – Cannot Pet or
                     Dominate Animals...160
         10.3.6.8   Can Do the Work Given – Can't Act in
                     Different Roles and Not Trustable....................160
         10.3.6.9   Can Give Lecture – Cannot Teach Various
                     Children According to Their Knowledge
                     and Intelligence Capacity 160
         10.3.6.10 Can Travel – Cannot Do a Solo Travel to
                     an Unplanned Location and Come Back........161
10.3.7   Solution and Research Direction 161

11. **Soft Cyborgs and Cyber Physical Systems by WT**................163
    11.1   Soft Robotics 163
    11.2   Cyborgs..164
           11.2.1   Hard Cyborg and Soft Cyborg..............................165
    11.3   Cyber Physical Systems...166

Foreword

Skin-Close Computing and Wearable Computing may be a strange title, but the book's contents will not be strange to those interested not only in wearable technology but also in assistive technology. An introduction of a wide range of recent human-centric technologies, such as COVID 19-protective clothing, sporting gadgets, firearm, health care (including for pets), etc., covering both hardware and software is provided. Futuristic wearables applied close to the skin are also discussed here. The author's knowledge in gloves and flex sensor technology has enabled him to cover these technologies with a wide range of applications in different domains. Brain–computer interface (BCI) and other assistive technologies are also covered in depth with the aim of helping those with severe physical disability. Security aspect is also dealt with signature verification and other recent biometrics such as plethysmography (PPG). Power requirement, which is an extremely important aspect of wearable computing, is dealt succinctly. Human–machine fusion, known better as cyborg technology in sci-fi movies, is also discussed. Overall, there are plenty of thought-provoking models and ideas of product development (some with in-depth technical know-hows) presented (some have won awards). The book would surely be of interest and *open the minds* of researchers to be *crazy* and think *out of the box* to develop technologies benefiting the humankind.

Dr. Palanniappan Ramaswamy,
Reader, & Head, Data Science Research Group,
University of Kent, UK

Preface

A keen interest in Biometrics, from the year 1998, took me to end up in registering with Brain Computer Interface (BCI) for my PhD research topic. BCI comprises many technologies, such as Signal Processing, Cognition, Evoked Potentials, Wearable Technology, Rehabilitation Engineering, and Assistive Technology. I found it interesting that all the above technologies have two dimensions of operations. I started it on the basic side to assist the disabled and found it interesting. I dared to venture into the next dimension of it and found that, when employed, it augments the commons to a supernatural level than other humans in performance. Many research works, papers, and presentations made me to swim deeper in the subject and revealed the great potential it has on a wide range of applications and business. Accumulated subject matter and the snowball effect of tools like Machine Learning and Deep Learning on to it makes me to look for a bulk avenue to deliver the entire knowledge bank since no keynote address or a technical talk satisfied me to handle the entire content.

Many areas I touched in this book are my personal research experiences and on-going projects. A few areas are underexplained due to their sensitivity to the on-going projects for military usage, but any person who has a keen interest on this subject can understand what I have written by reading it in between the lines.

My desire is to shift the focus of young students, innovators, and researchers to this potential area for their worldwide recognition and rewards by joining themselves into this research study abundant with ideas and applications. Wearable Technology is an inevitable component of the future systems like cyber physical systems, a refined arena of cybernetics, and will occupy every automation in the world soon.

I welcome questions and ideas from the readers and ready to do knowledge contribution to any of their work by suggestions if they are really into the business for the benefit and well-being of human lives.

Acknowledgements: Special Thanks

Thank you Triune God Almighty for letting me to complete this mission by giving me every good thing, starting from the publisher to the background music, talks, support of man, and machines.

I thank the chairman, managing directors, executive director, principals, and colleagues of Mahendra Educational Institutions for their continuous encouragements.

Thank you my dear research scholars for bearing my pressure and tight schedules on the experiments and documentations you did with me for your graduation and postgraduation degrees.

Author's Biography

Prof. Dr. Andrews Samraj is the dean and professor of Computer Science and Engineering working at Mahendra Engineering College, Nammakal, Tamil Nadu.

He is a renowned rehabilitation engineering scientist with cybernetics, assistive technology, and wearable computing as the area of specialization. His relentless service to the humanity can be categorized under science and engineering technology, human resource development to the nation and socio-economic development to the community. As a research supervisor in Information and Communication Engineering, of Anna University, Tamilnadu, he is presently supervising eight PhD candidates and Leading the R & D Team on various state-of-the-art Computer Engineering topics.

He has been recognized internationally as an Indian scientist who contributed immensely to the field of rehabilitation and security for the paralyzed, bedridden, physically challenged, and seriously ill patients in a non-invasive way of ability augmentation for the improvement of their quality of life. His innovations in this area won the NASSCOM Social Innovation Forum Award 2017,[1] the nation's highest award of this category. He received further awards from IBM in years 2005 and 2006 for his projects. He is the recipient of 'Distinguished Professor' award from AIRF and The best Engineering Teacher Award for the year 2017 from ISTE. His patent application for a biomedical engineering product was published in 2018. He has received various funding grants from India and abroad for his research projects, supported and recognized by agencies, including UGC and MSME. His outstanding contributions in these fields were recognized by many international reputed journals and international conferences. He has been organizing many international IEEE and other conferences as an active speaker as well as member of their technical and organizing committees. Recently, his invited lecture in the 106th Indian Science Congress at Punjab on ICT for nourishing the ethical behavior of Indian Youth was highly applauded by the nation's scientists, academicians, and student community. As a technical member, participant, presenter, session chair, organizing committee member, member of editorial, chief editor, and as a chair person his contributions always tend to a grant success. As a lead scientist in Wearable Computing, he carries out his research in signals, cybernetics, control, robotics, assistive technology, affective computing, and biomedical engineering and significantly contributes to the Indian research power. He is leading the AI & Deep learning research team in Mahendra Engineering College at zonal level in leadingindia.ai project. With his 20+ years of international experience, he authored and co-authored more than 100 research papers and has written several book chapters, as well as profuse magazine and newspaper articles on scientific and general subjects.

Apart from rehabilitation, his researches of Potential Fishing Zones of Indian coast to help fishermen community draw a good attention nationwide. He further contributed his findings to combat cyber warfare and cyber terrorism, F-INSAS model of Indian future soldier system, Brain–Machine Interface (BMI)-based communication system for locked-in patients, and security systems for disabled people. He is an academic advisor to National Cyber Safety and Security Standards (NCSSS), India.

As an innovative social and technological researcher, through a holistic and systemic approach, he found the relationship between the attitude and character of a child with the video games the child plays. Hence, he developed a system with meditainment and a new word coined by him called innotainment to heal, enhance, and direct the child's intellectual brain activity toward constructive, rather than destructive, paths by developing games with healing components inside.

His knowledge in linking up emerging behaviors and societal needs to innovative technological solutions leads him to conduct workshops, invited lectures, and presenting papers all over the country and abroad. His visits to foreign countries like Singapore, Syria, Sri Lanka, Indonesia, Israel, Jordon, Qatar, Cambodia, Bahrain, Thailand, Turkey, United Arab Emirates, Vietnam, and East and West Malaysia as a renowned scientist brought an awareness to the wearable technology field.

He held positions in Multimedia University in Malaysia, Asia Pacific Institute of Information Technology in Malaysia, VIT University, India, Kumaraguru college of Technology, India, and Stanes Motors Ltd., India. His excellence in teaching brought the President Appreciation award from Multimedia University Malaysia, and he won the IBM award for Best Final Year Student Projects for two consecutive years, 2005 and 2006, under his guidance. He has established a startup called Advanced Science and Technology Research Services for addressing the emerging demands of the digital community especially focused on the accessibility and assistive technology. He has produced many postgraduates through the practical way of hands on research and is continually guiding research scholars toward quality and outstanding research and PhD. He has been serving as an examiner for many doctoral candidates in computing and IT. He also served as a member of board of studies for many universities.

Apart from Science and Technology, he also contributes to the youth power building of our nation in various capacities, including career guidance. He diligently involved himself in NCC and got all 'A', 'B', and 'C' certificates and won the award of 'University out Standing cadet' from Bharathidasan

University in 1991. Later, he served as a dynamic Associate NCC Officer in Kumaraguru college of Technology, Coimbatore, in the year 1996–1997 as 2nd Lieutenant attached to 4 TN BN. He also pioneered the effective implementation of Blood Donors Club as its officer from inception in the college. He is widely traveling inside the state of Tamil Nadu and delivering Motivational and character-building lectures and workshops for school and college students. He has met more than 60,000 students in last few years for the above-mentioned purpose. He has been giving effective counseling and development training to numerous poor students free of cost to carry out their higher education. He served as an advisor to the International Student Society in Multimedia University for more than three years. He has organized and conducted international trips for students at Mahendra Engineering Colleges and took many teams of students to foreign countries like Malaysia, Singapore, and UAE. His motivational and technical talks in MMUMalaysia TV, Dharmapuri FM radio, and on Local Television channels in Salem received wide appreciation from the public.[2]

He has completed his postgraduate degrees in computer applications and computer science and engineering in India, and his PhD is from multimedia University, Malaysia, and the postdoctorate diploma is from World Scientific and Engineering Academy and Society (WSEAS) and Hellenic Navel Academy of Greece.[2] He has produced many scholars under his eminent guidance from various universities with PhD and masters by research degrees. He is hailing from an academic family, both of his parents and late grandparents were retired teachers from school service, worked diligently for the backward area students of Tamil Nadu.

Notes

1 http://nasscomfoundation.org/get-engaged/nasscom-social-innovation-forum/past-winners.html
2 www.andrewssamraj.com

1

Basic Wearable Computing Requirements and Advantages

A wearable computer is an idea hatched out of the continuing efforts to reduce the size and increase the mobility of computers. Though the wearable computers started with a backpack design and a hand-held monitor, there is no limitation now to the place of wearing it because of its small size and connectivity. The major challenge of wearable computers is its battery size and weight since it has to be carried all the time during operations.

The advantages of wearable computers are very obvious, such as the small size, mobility, handling both unconventional and conventional inputs, and convenience of ubiquitous operations.

The conversion of traditional computers from an office machine look to a combination of a more intimate, carry along, anytime, anywhere characteristic wearable brought it to the place where it is now. The computing concepts of input preparation from papers to tapes are all changed into a new generation of input devices called sensors. Thus, the office computing is changed into skin-close computing with all its advantages and thrills. Instead of blaming wearables for any disadvantages, the scientists of this field are taking it as a continuous challenge and are trying to remove, reform, and change any adverse effects into an advantage.

It was the fashion and trend of the late '90s for the businessmen, IT professionals, and students to carry a laptop bag. The advantages seen by this 'out of lab' experience were enormous, and later the use of laptops became unavoidable and came into common practice. Some companies immediately jumped into the production of palmtop computers, and it reached well among the young business people and executives. The touch-screen technology was effectively adopted in such machines and was greatly helpful in all operations, including the incorporation of mobile phones into it. But nowadays, the mobile phone has taken all these palmtops and allied devices into it and has expanded its scope with enormous application range.

The requirement of a mobile phone is a typical example of the requirement of a wearable computer now. The wearables, such as Bluetooth devices, remote devices like CCTV cameras, phone-to-phone connectivity, phone-to-device connectivity, and many mobile phones connecting into common apps, are the enhancements made to the common area between mobile computing and wearable computing.

DOI: 10.1201/9781003052906-1

The basic wearable computing requirements are not much different from a conventional mobile computer or a desktop, except in size, but include additional equipment to carry/attach them to the body of a person. Input, CPU, memory, output, and control units are very well present in wearable computing architecture but in a different form or shape that are ergonomically comfortable for a person.

The traditional input/output (I/O) devices are to be replaced with a real-time I/O device, such as sensors, to simplify the system design in the wearable computers that in turn made the process load less complex.

> In summary, wearables contain input sensors, camera, microphone, temperature/heat sensors, moisture sensors, bend sensors, shock sensors, eye trackers, etc.
>
> Its output devices are head-mounted display, speaker, microdisplay, flat panels, text-to-speech, tactile output, nonspeech auditory output, paper, and olfactory output (scent), etc., and a wearable can respond to its interface without a processor or processing.

1.1 Wearable Computing and Wearables

The first-generation wearable computers are to be literally carried on the body and clipped, hooked to the body and dress. Wires also need to be plugged for connections. As a new generation with the advent of skin-close sensors (tattooed/sticked/skin touchable) it was made simple as wearable to wrists, fingers, forearms, head, ears, chest, and every human body part. ZYPAD is a typical example of a wearable computing PDA braced to the hand and called a wrist-worn PC, weighing less than 300 gms. It uses a CPU AU 1100 @400MHz with Linux kernel 2.6 and windows CE 5.0 as operating system of choice. It inherits most characteristics of a laptop, such as automatic standby mode, dead reckoning, GPS connections, Bluetooth, IrDA, and Wi-Fi.

The difference between wearable computing and wearables is wide and thin. Wide in the case of passive and traditional wearable material/objects and thin in the case of smart devices, input medium, and data acquiring devices that are in the form of wearables. Technically, the traditional wearables like ornaments are not relevant to the technology unless they are modified, embedded with some components of technical purpose. Hence, to the context of the wearable technology (WT), 'wearable' directly refers to the devices that work along with the technology for quantification, communication, and processing purposes in WT.

The thin difference mentioned between wearable computers and wearables is similar to the Processing Elements (PEs) and a processor (µP). PEs are limited to homogeneous operations, predefined I/O functions, and recursive operations. But a processor is independent, versatile, and heterogeneous in operations. Similarly, the wearables can be defined as a part of wearable computers (sometimes as a whole) that passes the necessary output information, raw or processed, and can be used, interpreted by other parts of the application technology framework. Many companies like Fossil, Sony, Seiko, Timex, Hitachi, and Panasonic also came into this wearable computing market and contributed to their portions. Google ventured with a complete paradigm shift in wearable computing from 2013 onwards. At this point, the specialized smart devices like earbud wearables were brought into market by LG and IRIVER and shifted the wearable computing from all-purpose wearable computing to special-purpose wearables. Another example is the WSS1000 barcode scanner.

The WT outpaces the traditional, less flexible, and heavier equipment into an easy, lightweight, and fashion device. The fitness tracker headsets that play music and listen to what the user speaks to initiate an activity, advanced sports watches that record and replay the statistics of a player's performance, augmented reality headsets that give not only an entertainment but also tutorial based on the inputs given. These are all the examples of wearables that adopt a little processing which they are capable of performing.

Wearable devices are mostly portable, but not all the portable devices are wearable. There is a slight difference between them. Wearable devices are distinctive from portable devices by letting hands-free interaction, or minimal use of hands, when using the device. This is achieved by the devices when they are actually worn on the body. Examples of such devices are head-mounted devices, wristbands, vests, belts, shoes, etc. On the other hand, portable devices are always found to be compact, lightweight, and they can be easily carried but not worn by the user and definitely require constant hand interaction. Examples of such devices are tactile displays, electronic canes, mobile phones, laptop computers, etc.

An advanced version of wearables has taken up the form of stickers and patches covering a good range of applications, including self-protection for women. A recently developed sensor patch in MIT, which could be attached to the dress of a person, will learn the regular undressing pattern and raise alarm when forcible undressing takes place. It works on both a passive and active mode, which helps the user to initiate creating alarms and tele-contact establishment by pressing the sensor in the former stage and reads the signals from the external environment and acts accordingly in the latter stage.

A typical example of the destructive side of the WT is the suicide bombers who are being a challenge to WT scientists to find a wearable device which can raise alarm in the presence of such bomb carriers. A novel technology must be infused to prevent or detect the misuse of this technology (Figure 1.1). Many suicide bomb attacks in the form of wearables were reported in nearly 38 countries, including India, Afghanistan, Israel, Sri Lanka, Philippines,

FIGURE 1.1
Wearables for peace and welfare

and the United States. Thinking in that direction, even the need for a device which can alert people if their social distancing is violated is felt essential during the epidemic diseases like the recent Covid-19 virus outburst started at Wuhan, China.

1.2 Wearable Types

Wearables are wide in variety and applications, and though the place of wearing and the shape and size differ, the sensing and operations are alike; hence, it is difficult to separate the wearables with a clear boundary. Hence, even if classified they have overlaps in operations and applications.

At the same time, wearables can be classified as external wearable and internal wearable. Some external wearables need skin contacts and some do not. Some wearables are passive and some are active all the time of their operation. Examples of an internal wearable are a pacemaker and a smart pacemaker.

Just like the ethical emphasis of 'Responsible Robotics', a foundation that responds to the adverse effects of automation on human nature of living, those who are interested in the area of WT have to be 'accountable and responsible' in the laws and policy associated with the design, construction, and usage of wearable devices. To the point, the WT has to be created and deployed for creative, assistive, and noble purposes that benefit and are for the well-being of every human. Another important reason is to help, combat, or support humans in the usage against any adverse and destructive technology that puts human in danger. This purpose will be discussed in this book in detail in the section discussing soft cyborg creation.

1.3 Wearable Computer Examples

As mentioned in the previous section, the form of wearable computers has taken new transformations. The following few examples reveal the depth of wearable computing utility and its flexibility to get infused into the devices of different shapes and functioning. Generally, people assume that a wearable computer would look like a palmtop computer that can be worn like a watch on the wrist or a coat that have many touch sensor switches and a processor chip embedded somewhere in the same coat, etc.

The wearable computers are not expected to be like the systematic full personal computers (laptops and desktops) used for multiutility. The ultimate aim of wearable computers is very focused and customized. The wearables are placed in a different seam to address specific applications effectively and sometime collectively. We can say that the wearables are a form of cyber physical systems if connected with the integrated output devices. A few typical examples are as follows:

a. A wearable PrioVR accessory, worth of $400, for virtual reality gaming that connects the player and the computer is available in the market.

   Here the inputs according to the gaming scenario are taken from the hand and head movements of the player and are implemented in game moves.

b. A palm-size device with OLED display is capable of recording all activities like steps, distance, calorie intake, heart rates and delivers it to other digital devices for display and analysis.

   It also contains some guidance program for exercises to achieve wellness targets effectively.

c. A wearable camera enabled with a 4G LTE protocol that functions without a SIM or antenna for streaming videos on the go of the user wearing it.

   A very essential and versatile equipment useful for media, police, and other entertainment industries.

FIGURE 1.2
Halo – to maintain work or social distance, Apollo – to release stress

 d. A neck belt with a 3D accelerometer sensor for monitoring the pet activities costs $99. This records all the active and sedentary timings, repetitions, intervals, etc.

 e. Many other similar devices in the form of goggles, watches, gloves, etc. for recording the data in digital form for further analysis and reporting.

In general, the wearable computing has increased the health consciousness, safety, and performance enhancements among the users. Recently wearables are in the smart business that utilizes the opportunity of epidemic, releasing lot of applications to protect people from infectious diseases. One such example is Halo wearable wrist band that helps to maintain a good work or social distance released by Proxxi of Canada. Halo is very suitable for schools, colleges, and factories since Halo produces a beep sound and draws the attention of wearers denoting that another halo band is within 2 meters or 6 feet (Figure 1.2).

Wearables are brought as close as the skin to ensure uninterrupted utility and service since it is the next close alternative to an invasive system which is not preferred always. A thin skin-like wearable circuit that can be sticked on the skin has already emerged, which brought the wearable computing to a skin-close computing by forming a second skin and track all details which a user can do through a data glove. A pioneering product called Pysio-Power heal will be discussed in the subsequent chapters.

2

Ergonomics and Its Benefits

Starting from its creation, the human race of this world has come across many eras from the past. Every phase of human era is different from each other in many ways due to the changes that have happened on the surface of earth and the atmosphere. Apart from the nature and related environment, the empires, powers, wars, and involvements in mass missions have also impacted the human life span. The adaptability to survival, struggle to protect life, was also added to the list. All these transformations have been directly influencing the age and comfort of the humans in both gender, especially the social reaction toward the ever-growing machine interaction has been carried out by humans in all colors from the ages past. Whenever a machine is created, the continuous refining of the same also coexists, and there is always a significant share for ergonomics.

2.1 Definition

Ergonomics is the way in which things, workplaces, products, and systems are designed and made to suit naturally and comfortably without any adjustments to physical postures, etc. of the people who use them. Hence, ergonomics is sometimes called human factor design that fits tasks and equipment to the user perfectly.

2.2 Why Ergonomics

The word ergonomics, and the phrase 'ergonomically designed', is heard when there is a talk about seats, sofas, cars, etc. But ergonomics is not only for seating comfort but also for everything that people use, including their mobile phones. The sportswear, such as shoes, is a great example for a wearable that needs ergonomic design. The new synonymous name for ergonomics is 'human factors'.

The focus of ergonomics is on fulfilling the objectives of safe productivity and occupational health and safety.

DOI: 10.1201/9781003052906-2

The benefit of ergonomics is not only comfort, but it plays a major and crucial role in avoiding wastage and accidents. Even a flooring, or steps design to, tool boxes screw driver handles, steering to aircraft cockpits are following ergonomics for comfort, easy, safety, and efficiency. Many organizations are working in this research area to bring out the enhancements and fresh designs to make the life of humans comfortable. The field of ergonomics started its applications 70 years before from now and is still growing by encompassing different fields like orthopometry, biomechanics, environmental physics, applied psychology, social psychology, user interface design, and many more into it.

2.3 Work-Related Injuries and Causes of Sickness

In a famous hospital at Tiruchirappalli, in India, a sudden occurrence of repeated cases of hand-burn injuries in women involved in cooking was reported during the period of early 2000s. All cases reported more or less similar reasons for getting their skin burnt by hot vessels on the stove and few in direct flame. Later, the reason for all these burnt cases was found to be the new four-burner stove, which was introduced in market and was available to all upper middle class at that time. The judgment of women went wrong many times in handling the vessels while all the four burners are working and that leads to the hand burnt. Later, the three-burner stove seemed idle and occupied the maximum sales due to this ergonomical failure of four-burner gas stoves. The ergonomics is not only in the shape but also in the placement and frame.

The problem of aging was a great topic of research before ten years, and still people are touching it when necessary, while suggesting solutions like telemedicine, fall detection, smart home, etc.; the strong requirement of ergonomics and wearables is unavoidable to deal with the problem of aged people. When it is essential for aged population, it goes without saying in the case of disabled.

The cost of work-related injuries and illness in developing countries are raising high due to the non-ergonomic workplaces. Even in developed countries like Australia, according to 'safe work Australia' the expenses on this are estimated to 60 billion dollars per annum.

In the long-omitted area of agriculture, farmers usually complained of lower back pain and also suffering due to lack of ergonomically designed equipment. It is not only for farmers but also for the office workers, site engineers, etc., to experience repetitive strain injuries as well as musculoskeletal disorders that may lead to disability over time.

Apart from physical ergonomics, cognitive ergonomics and organizational ergonomics are the important branches of ergonomics that play their share in human factor engineering.

There is more than one reason for bringing the ergonomics into wearables. The wrong-size shoe or a worn-out shoe causes severe damage to heel, knee, hip joint, and later the nervous system, based on the duration of usage.

2.4 Ergonomic Importance of the Present

A researcher, Vimala, worked on her research investigation during the early 2000s on ergonomic problems of using button mobile phones to send SMS. The effect of hand anthropometry on Short Message Service (SMS) was studied using structured questionnaire interviews with hundred and ten (110) subjects, aged between 17 and 25 years. The observations from the outcome of this survey revealed many interesting factors. The size of the hand was measured to assess its effect on mobile phone design factors and satisfaction. The thumb circumference and length were also measured separately for keypad design factors. The subjects with small hand sizes were expressing their satisfaction with mobile phone dimensions than the subjects with large hand sizes. The thumb circumference significantly affects users' ease of use and satisfaction on the key size and space between keys whereas thumb length significantly affects keypad layout satisfaction. According to the survey, both the thumb circumference and length significantly correlate negatively with the corresponding keypad design factors. In summary of the findings, it is confirmed that the hand anthropometry does affect the messaging satisfaction of the users. These findings were very much useful to mobile phone designers who could look into the possibility of designing customized mobile phones that cater to large-hand and thumb-sized users, so as to increase their subjective satisfaction. Another survey similar to the above was conducted to find the effects of hand-size variations and gender on mobile phone users' texting satisfaction in which users were investigated using a fresh structured questionnaire. Interviews with another 110 subjects (18–23 years old) were conducted. Focus of this survey was on text-entry factors, such as speed, learnability, simplicity, navigation, and special characters. The results of the survey showed a significant gender effect on speed and special characters' selection, with females being more satisfied than males. Again the hand-size effect is significant for speed, special character selections, and navigation with smaller hand-size subjects being more satisfied than subjects with larger hand size. Overall texting satisfaction among subjects with smaller hand size was more than the subjects with larger hand size, regardless of their gender. The researchers in their recommendation suggested to introduce an improved or new text entry mechanism which in turn would increase texting satisfaction, regardless of their gender and hand sizes. Since the outcome of the survey confirms that hand-size variations and gender affect users' texting satisfaction, the importance of designing the hand phones based on ergonomics is confirmed.

This study focused on mobile phone keypad design factors, namely, key size, shape, texture, space between keys, layout, and keypad simplicity. Females were found to be more satisfied with the key size and space between keys, whereas males are more satisfied with key shape. Users with smaller hands and thumbs were found to be more satisfied with key size and space between keys compared to those with larger hands and thumbs. One of the recommended improvements was to have larger keys with more space between them. Results obtained can be used by mobile phone designers to design customized mobile phones, for example, mobile phones that suit users with larger hands and thumbs, especially males.

Another interesting study of ergonomics proved that it works beyond hand-held devices in making people comfortable. An investigation was made by these researchers for an alternative approach to solve cooling problems in two distinctive shopping malls in Malaysia. In both the cases, an investigation was initiated with an idea to capture a general opinion of poor indoor cooling suffered by the shop owners, business operators, and other patrons. Details about ergonomics methods were collected through unstructured interviews and direct observations (DOs) to obtain information on the major complaints from the stake holders. A technical inspection by visual methods into the air handlers and throughout the air ducts was also done. The response of the business operators was tested utilizing the subjective assessments (SAs). A detailed budget analysis, on both current and archival data pertaining to the maintenance service of air handler, was retrieved for this purpose. A variety of ergonomic interventions were incorporated into the existing setup by installing, modifying, adjusting, or replacing the system by applying basic sciences for rectification work. The previous design specifications were altered to fit into the new design. Systematic follow-up interviews and studies were done after the new installations were in place, using the same methods of data analytics like DOs, SAs, using current and archival data. The results of such analytics show the effectiveness of the interventions suggested. It is further realized that the installations were cost-effective and the new designs improved human comfort level through effective heat removal from the same air-conditioned space. There were also cost savings found due to the maintenance done to the new coil assemblies.

2.5 Ergonomics in Wearables

Starting from dress, a bangle, or any other ornament, there is a conventional comfort in wearing them and they never hinder the day-to-day activities. The wearable computing is also greatly grounded on the ease and comfort

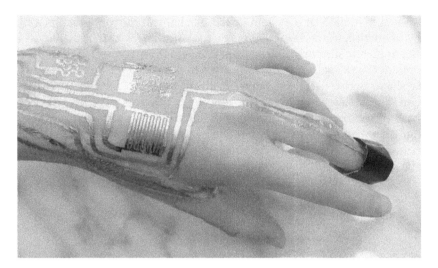

FIGURE 2.1
Wearable circuits drawn (printed) on the skin
(Source: The week magazine November 17, 2020 12:25 IST)

of ergonomics. The human tendency to always have a 'free hand' fuels the quest for wearable computers. The development of projects like F-INSAS (Future infantry soldier as a system) was a great propellant to the wearable computing devices. Ubiquitous and pervasive computing concepts included the WT into their domain, growing along with it. As a result of continuous research and development in the field of WT, the fabrication of flexible and stretchable electronics has joined the group of wearables as an epidermal electronic device. It adopts a novel way of sensing technology that is printed directly on the skin. These skin-close wearables act as a temporary/permanent access control keys, fitness trackers, and health gauges and can be attached to smartphones or other gadgets through an app (Figure 2.1).

2.6 Ergonomics and Biometrics

The ergonomics includes measures and dimensions of human features that adopt the wearables. This nature helps in biometrics which measures the uniqueness of human features that are tangible to identify them. A wide range of biometric applications started getting into wearables that incorporate these biometric features for safety and security of the user and the applications.

A novel invention of a system called mechanical mirror, which is a huge array of small and identical square pieces of different materials like wood, metal, or plastic supported by sensors and actuators, connected to their stepper motors on a wall or stand. It senses the presence of humans based on their shape and reflects the same by tilting the exact number of pieces to resemble the design while other chips don't tilt. This may be artistic, but researchers working under human surveillance without a camera or gesture recognition may feel it as a good human recognition system alternative to a camera-based system. So human shape or any other shape works as an ergonomic input to the mechanical mirror to perform some activity which can be further utilized for many other purposes of security and identification.

3

Applications of Wearable Technology

Wearables are of various types according to their nature and applications. Here, you can see some examples and their use. These common wearables are potential to house the electronic components inside and get transformed into smart wearable. Once it becomes a smart wearable, it gets its transformation as a system and can be a component of any larger system for any application of a higher level.

Basic wearable: (Dresses). The first wearable was invented by the first man on the earth who created dresses out of the fig leaves. The basic need of dresses is for protection and dignity. Next comes the style and fashion for beauty and attraction. Wearing LED lights on dresses by the dancers is common nowadays. Now the wearable, washable, and flexible electronic components that can be directly incorporated in the fabrics are developed.

Since it is the very first important wearable, many sensor-based wearable devices are incorporated in the dresses so that the wearables are well absorbed with natural tendency of ease. This leads to the opening of entirely new applications by customizing textiles into functional energy-storage elements, harvesting energy from body and storing energy to supply to other applications like the Internet of Things. Even the wearable dress application can make a man invisible!

3.1 Wearable Types and Applications

The applications of wearable technology in wide range of fields started very early since the humans found it convenient, comfortable, and concealed.

3.1.1 Fashion Wearable (Ornaments, Tattoos)

Differently named as Smart Jewelry, Ornamental type wearable devices encompass a great range of applications, including protection against assault and rape. Many IOT-based smart ornaments are being developed every day. Identifying the opportunities in this potential area of business, an Indian jewelry firm called PC Jeweler Limited made an agreement with a US-based watch production company called Martian Watches and ventured into

DOI: 10.1201/9781003052906-3

FIGURE 3.1
Smart ring that incorporated a tracker, alarm, and a notifier

making 'smart jewelry' products. Pen drives are now seen inside the pendants of the neck chains, finger rings, and bracelets (Figure 3.1).

3.1.2 Sports Gear (Shoes, Knee Cap, etc.)

Apart from foot and knee cover, cushion, the new technology in wearable balances the person wearing it and also measures all the movements, pressure, sweat, etc. if fitted with appropriate sensors. The output analysis is not only for performance and improvement but also for many other tracking applications.

3.1.3 Comfort Wearables (Pads, Elastic Bands, Thermal Comfort Wearables)

Any outdoor or indoor activity that requires flexibility, convenience, and comfort has to be supported with appropriate wearables. Physiological signals like skin temperature, heart rate, etc., can be collected and a thermal control circuit works to make the wearer comfortable. A recent wearable in market, made of elastic material, which goes with the skin color, raises the height of the user to few inches and goes inside the socks comfortably. It gives a good cushion effect to heals and raises the height of the person.

3.1.4 Wearable for Childcare (Baby Carriers)

Baby carriers on front or back were created as a modification to prams. There are many things wearables can do by taking part in child monitoring as a part of baby carriers.

3.1.5 Wearables for Recreation (Tokens, Dollars, Wristbands, and Tattoos)

Amusement parks like Disney Land, Ferrari World, etc., did a survey of customers' feelings and found the visitors are not happy in long queues, waiting, food order, carrying wallets, keys, etc., during the time of their entertaining moments inside the parks. So Disney launched a digital initiative called MyMagic+. This MyMagic+ combines a website, a mobile application, and a wearable wristband to upgrade the visitor's experience and satisfaction. The bands give them wings to skip queues, fasten their rounds, make the food order in restaurants easy, and even have access to their resort rooms. Same has been already implemented in many schools and colleges for the use of students to spend for their canteen orders, book stores, and other amenities.

3.1.6 Wearable for Road Safety (Helmet, Seat Belts)

No one can deny the greatest benefits of these life savers. Meanwhile, the cars will not be warning you anymore to wear the seat belts but the seat belts will automatically embrace your body soon in future. Smart helmets give much guidance during day and night for umpteen purposes in on- and off-road applications.

3.1.7 Wearable for Life Safety (Life Jackets, Anti-Bomb Suit, Biomedical Suits)

How can one deny the benefits of the greatest protection package in these jackets and variants for multiple purposes? Recent virus, Covid-19, originated from Wuhan, China showed doctors and medical workers in a new form of appearance with such lifesaving dress.

3.1.8 Operational Wearable (Gun Holders, Tool Holders, Armor Belts)

Many comfortable gun, ammunition, and pistol holders with both hidden and ceremonial display are a part of people working in armed forces and recreational shooters. Workshop men and carpenters wear a dress with many pouches to hold their tools comfortably; hence, the pilots of fighter jets followed the same to accommodate their special needs of performance.

3.1.9 Camouflaging Wearable (Nets, Leafy, Grassy Suits, Gilly Suits, and Helmets for Snippers)

Starting from face painting and camouflaging uniforms, chef hat helmets, helmets with grass-like elements to a total 'invisible man' concept dresses are all using the wearable technology nowadays (Figure 3.2).

(a)

(b)

FIGURE 3.2
(a) F-INSAS; (b) Indian Para army in camouflaged attire during the field training

3.1.10 Wearable for Work Place/Manufacturing/Plantations/Farming/ Cooking (Apron, Gloves, Mitten Cap)

For maintaining hygeine in food processing from farm to factory, people use specific dresses, and now these dresses are coming with many wearable applications like automatic counters to do the work of counting to avoid human error, quality control, etc.

3.1.11 Wearable to Carry Weapon (Rocket Launcher Belt, Grenade Hooks, Bombs, Suicide Bombs)

Undetectable to metal detectors, people started developing plastic or rubber bombs and other detonating devices. It's a good challenge to find them since conventional metal detectors fail to find them. The same WT helps them to design such elements and also helps the armed forces and security personnel to identify such nonmetallic items of hazardous nature by the X-ray scatter tomography. Explosives like rubber bullets, nitrate esters, nitroaromatics, nitramines, chlorates, peroxides, and drugs like cocaine, THC, ketamine, opiates, and methamphetamine are detectable using new technology which is now transformed into the wearable detectors.

3.1.12 Wearable for Protection Against Climate (Puffer Jacket, Warmers, Sweaters)

There are special materials available from ages back which we use for protecting people from severe cold weather. Reindeer skin to geese feather jackets of traditional methods to the latest thermo control jackets inbuilt with water heaters are in market. But if the heater-based jackets are used, they are more detectable by a thermal camera, which is to be seriously addressed in the case of security personnel working under the surveillance of enemy. Smart wearables to address this issue are under dynamic research.

3.1.13 Wearable for Protection Against Water (Swim Suits, Dive Suits, Pressure Resistive Suits)

Flexibility, convenience, protection from UV rays, protection from water pressure, protection from undesirable temperatures are a few benefits the modern dive suits provide to the divers in different weather conditions of underwater diving. Dry suits, wet suits, semi dry suits, hot water suits, atmospheric diving suits, and dive skins are few examples of dive suits used under different conditions and purposes of diving.

3.1.14 Wearable for Protection Against High Pressure/Low Pressure (Suits that are Designed for Divers and Astronauts)

The dive suits of deep-sea divers protect their organs, especially lungs, from the dangers of hydro static pressure of the deep seawaters. The pressure increases in all parts of body at a rate of 0.445 psi per foot depth of seawater. Immersion of the human body in seawater affects the blood circulation, renal system, fluid balance, and breathing. The astronauts wear more sophisticated suits during their space operations.

Unlike the popular videos of astronauts swimming in regular dresses inside the space ships, sometimes they need the suits even inside the spacecrafts. Recompression dresses are also used to treat the disorders caused due to the compressions in the divers and astronauts.

3.1.15 Wearable for Fire Safety (Fire-Resistant Suits)

A fire protection suit called as proximity suit is a dress specifically fabricated for the protection and operation of firefighters and volcanologist. This proximity suit is capable of withstanding its integrity and high temperatures during the firefighting and rescue operations. Former versions of this dress were made of asbestos fabric, but the recent advancements use vacuum-deposited aluminized materials. These suits incorporate boots, gloves, and helmets as their components along with the full body covering. A breathing facility has to be included in these suits since the face shield is made out of heat-resistant material which can reflect at least 950° C. Even a radiant heat temperature of 150° to 2000° C also cannot harm the person wearing this dress, and it provides protection against direct flame. The fire-retardant screens were made up of 100% Aramide lining laminated windproof, but breathable climate membrane is made of polyurethane or cotton material coated with calcium salts.

3.1.16 Wearable for Protection Against Vacuum/Cosmic Rays/Heat (Astronaut Suits)

NASA designed and produced space suits and updated its design and utilities frequently on feedback. The new space suits produced by space-X gained a five-star applause from NASA scientists since the latter surpasses all the comfort, appearance, flexibility, safety, and fashion of the former. The sleek suit very close to the formal outfits on earth replaces the old pumpkin-shaped dresses of the astronauts like Neil A. Armstrong was wearing on his expedition to moon. In 1969, he took a 'small step' on moon out of the spacecraft on which he travelled to moon and became the first human adventurer. In space, the astronauts face heat and cold to the extremes. The space suit should be capable of protecting the astronauts from such temperature extremes since the radiation causes the temperatures at sunlight side to 300°C and shaded side to -267°C. The space suit of astronauts walking in space is just another spacecraft and possesses all necessary instruments in it under the four-layer architecture. Any disturbance in the supply of oxygen and cold air through tubes called umbilical to the astronauts will cause the brain death and failure of organs in minutes. Similar to astronauts, the fighter jet pilots who cruise in higher altitude also wear tight-fitting suits to make them feel comfortable atmospheric pressure similar to the same they feel on earth.

3.1.17 Wearable for Protection Against Chemical (Anticorrosive Suits)

There are 100s of options available to combat people from the chemical exposure. The reason is the plenty of hazardous chemicals used. The anti-chemical suits that are airtight may cause raising heat inside the suit that is dangerous to human body. There are toxic chemicals, explosive chemicals, irritant and corrosive chemicals, and flammable chemicals that need different level of safety suits. Similarly, the difficulty level of protective action rises when the chemical that is dealt changes from solid to liquid and liquid to gas. Companies like DuPont™ claim that their use of a versatile material Tyvek® has the positive remarks of durability and comfort, providing an excellent barrier against dangerous fine particles and chemicals.

The Tyvek material is unique in protection built in by the fabric itself, and no films or laminates are included as conventional suits used to avoid leaks from stitch line holes. Whether performing asbestos abatement or installing insulation.

DuPont™ Tyvek® is a versatile material that is both durable and comfortable, providing an excellent barrier against dangerous fine particles and chemicals.

A unique nonwoven material with protection built in to the fabric itself, Tyvek® has no films or laminates that can abrade or wear away over time. The whether performing asbestos abatement or installing insulation were used in the earlier years, but they are prone to abrade or wear away over time. A self-protecting or neutralizing material dispensing system that may be attached to the wearable suit and can be operated during emergency is a possible enhancement to it.

The industries in which the protective suits are used include Military and Law Enforcement, Firefighter Protection and Emergency Response, General Manufacturing Protection, Controlled Environments & Cleanroom Protection, Automotive Protection, Oil & Gas Protection, Electrical/ Utility Protection, Pharmaceutical & Laboratory Protection, Agricultural Protection, Decontamination and Remediation Protection, Food Processing Protection, Transportation Protection, Mining Protection, Sanitation Worker Protection, Construction & Maintenance Protection, and Asbestos Removal Protection.

3.1.18 Wearable for Protection Against Germs (Masks to Protect From Virus like COVID 19, or VD, STD)

The medical personnel are always under biological threats when they meet patients for treatment, but in year 2020 the same levels of threat and protection were spread to common people. We have heard about biological weapons and understood that now there is no separate missile or bomb required to spread it. An attack called infection may happen to any human by exposures to blood and blood-borne pathogens, during surgery, accident, or transmission. Hospital workers or government teams require wearables for pandemic preparedness and response. Police and medical forensic personnel need protection against such microorganisms during crime scene investigation and cleanup. A good example for such cleanup and bioterrorism incident is the cleaning of KL international Airport-2 after the incident of Nerve-VX chemical being used against Kim Jong Nam the half-brother of North Korean premier Kim Jong Un in the year 2017.

The wearables used by the prison authorities during the cell extraction procedure help in the forcible removal of a dangerous prisoner from a cell by a tactical team armed with less-lethal weapons like Tasers, pepper spray, and stun shields. They need special wearables to protect themselves from any form of chemical or other attacks from the prisoner, since even when carefully conducted, cell extractions carry lethal risks.

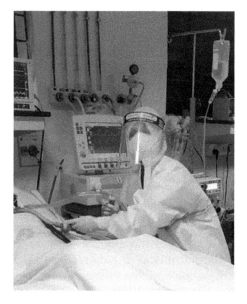

FIGURE 3.3
PPE suits for medical personnel when treating patients with contagious deceases

3.1.19 Wearable for Protection Against Animals/Fishes (Wire Mesh Swim Suits)

There is a dive suit called Neptunic C Suit, made with steel mesh, titanium, and hybrid laminates, which as a reinforcement can withstand shark bites. The idea of an anti-shark bite suit was in attempt from early 1980s. An Australian shark expert Ron and Valerie Taylor had an idea of making such a suit with steel mesh from the inspiration they got from the wire mesh gloves for butchers used to protect their hands during the meat cutting.

3.1.20 Wearable for Advertisements (Fancy Add-Ons)

Wearables are a viable medium to do very intimate advertisements since the user has the wearable always with them. Many new ideas of linking emotion with buying behavior result in the identification of the desire of purchase and its pattern. Sending advertisements to wearables just like the ads to FB and YouTube is also possible without violating privacy policy. Wearables enhance business to a great extent by supporting tracking, analysis, and model making. People use fancy dress wearables with advertisement words on their dresses to attract the attention of the buyers. Portraying as animals and cartoon characters for advertisements is still popular. Recently, advertisement workers are carrying glowing sign boards on their back and night walking in streets to do advertisement.

3.1.21 Wearables for Magic (Magnetic, Covering, Especially Stitched Garments)

Using different kinds of visible and invisible wearable gadgets can make magic. A magician can switch on or off the candles by the moves of his hand near the candlestand, thus closing or opening the circuit by the electrical induction. Any metal thing can be hidden or attracted, pulled or pushed by the magnetic wearables.

3.1.22 Wearable for Communication (Gloves Like 5DT)

The communication between special children, disabled adults, and patients and their aides or caretakers are limited and complex due to the extremely complicated disability of patients. A work done by our team on a clinical based patient monitoring used a wearable data glove to read implicit communications of such underprivileged subjects carefully from their gestures to recognize their need. The precision of such system should be of high standard to eliminate all misinterpretation that may complicate the communication to disaster. The wearable system is made in such a way that it can interpret the implied communication precisely to the caregivers or to any automated support device. The simple form of tangible hand movements are used for this purpose. This system leads to a simpler and faster communication than the sign language system which uses the same wearable data glove to establish an implicit communication and also improves the system by finding appropriate gestures for the better use.

3.1.23 Wearables for Wellness (Stress Busters)

Apollo, a wearable band introduced by Apollo Neuroscience, is the first in-market device that helps the body of users to adapt to stress. It is clinically tested and approved. It gives the user a good mind concentration by facilitating focus and even sound sleep. By the good rest, it energizes the body of the users and many more added health benefits. Apollo is different from other wearable bands like trackers and empowers the users by delivering gentle vibration waves and restores the mood and energy by massage and blood flow regulation.

Even artificial neurons to aid the cure from chronic diseases and wearable ECG sensors for 24X7 monitoring care to heart patients are already in use. Smart bras in market to monitor stress and discourage emotional eating are available.

3.1.24 Wearable for Communication with Vehicles (Gloves and Microphones)

In yester years' Spiderman cartoons, the Spiderman calls his cylinder-shaped helicars by his voice-recognition systems, which was not at all available in those days. Everyone considered it as a magic or fiction until we saw systems that respond to voice by recognizing it. Later, we developed automobiles and even aircrafts that can identify humans by voice, vision, and now by wearables. It enables a keyless, touchless entry into vehicles and incorporates many parameter settings.

3.1.25 Wearable for Managing Vision Impairments (Spectacles, Lenses, AR Smart Glasses)

Projects are successful in providing artificial vision through camera-based devices or wearables to vision impaired and to everyone at night or in places like caves and mines without light.

3.1.26 Wearable for Communication with Devices/Life Support (Gesture, Movement Capturing Gadgets)

Motion-control devices, wearable cameras, and gesture-recognizing devices are working on image- or video-based inputs; at the same time a wearable-based input makes it more simple, easy, and secure. The security part is more important to communicate with very sensitive devices – transactions of funds which need passwords or PINs. Wearables enable the security process, cancellation, and rewiring.

3.1.27 Wearable for Communication with Robots/Cyborgs (Gloves, Rings, with Embedded Electrodes, Lights, Sensors)

A verbal communication is not ideal or possible in all situations and by all persons. In order to use the wearable for communication, the gloves and rings are very much useful. They interpret signs and reproduce a sign-language communication into verbal and vice versa. We will see more on this in Chapter 11. It is possible to recognize persons and distinguish them without a camera eye for Robots using sensor technology through their gestures and dimensions.

3.1.28 Wearable in Warfare (Bulletproof Suits, Metal-Plated Suits, Proximity Alarms, Weapon Sensors)

Many projects and research by Indian Defense Research and Development Organization DRDO and other agencies are in progress in the military sector on wearables. Starting from a basic cooling jacket, which uses water to bring down the wearer's body temperature, to invisible attires, Proximity silent alarm dresses and hidden weapon-sensing dresses are notable among them.

3.1.29 Wearable for Pets

In Israel, some researchers found a neckband for dogs and cats to monitor the health conditions and intervention. Animals cannot wear an oximeter like humans since the fur on their body never allow a proper contact and function. So researchers ended up with this kind of innovative neckband to measure and pass the information about the body temperature, heart rate variability, and other major sensible body conditions for the owners of the pets to initiate intervention. When the body temperature goes up, they can switch on the air conditioner to create a pleasant environment for the animals. Similarly, the HRV will tell us the pain or other stress encountered by the animals so that we can initiate inspection of the animals to give them treatment. There are wearables for the protection and identification of pet activities too.

3.2 Development of Wearables

In India, the earliest known noteworthy historical incident with the wearable technology warfare happened on 10th November, 1659. It was the usage of tiger claws (Bagh Nakh) designed to fit over knuckles by the Maratha leader

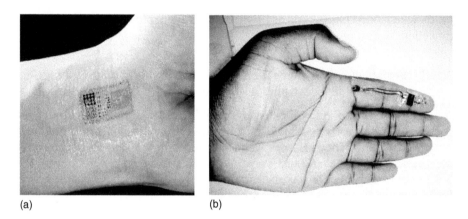

(a)   (b)

FIGURE 3.4
The latest wearable stickers on skin with multiple sensors

Shivaji to kill Afzalkhan, the general of Bijapur Sultanate and a commander of the Mughal army, in a historical encounter between them at Javali, located at the foot of Pratpgad fort, Maharashtra, India.

Professor Steve Mann, from the University of Toronto, started the wearable device-based computations in the decade of 1980–90. He designed a back-pack mounted CPU, a 6502-based multimedia computer with the capability of processing texts, graphics, video, and multimedia. Later, he started to incorporate webcams etc. and started recording events and regular activities of day-to-day life and calling it as life logging.

Smart phones, smart watches, welding helmets, Google Glass, health-care systems, and electronic textiles are also the components of wearable technology nowadays (Figure 3.4).

4

Simple User Interface Design for Wearable Computing and Wearable Technology

The aim of wearable computing is ultimately to provide the user with the freedom of mobility. When mobility is essential, it has to be easy and hassle free. So, the weight considerations and optimal use of trustworthy devices are given high priority while designing the wearable computing devices.

The basic things that have to be remembered in the design of wearables are as follows (Figure 4.1):

- Short ranged
- Simple design
- Low power consumption
- Self-configuration
- Restriction to the user
- Highly essential security

A typical virtual keyboard application with a wearable finger typing detection is a fortune for common public and their inclusive society. It is as simple as depicted in Figure 4.2 and easy to fabricate as we desire.

4.1 Construction of a Hand Glove with Flex Sensors

The hand glove is a widely used wearable for diversified experiments, and you can design it with the guiding diagram given in Figure 4.3.

1. Arduino Uno – open-source microcontroller
2. Battery – 3.7 v lithium-ion rechargeable battery
3. Flux sensor – input signal from the user
4. RF transmitter – wireless Signal transmitter from one place to another place

DOI: 10.1201/9781003052906-4

Wearable Computing Technology

WT Block Diagram:

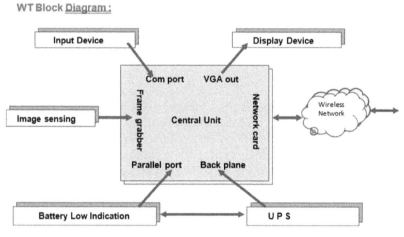

FIGURE 4.1
Block diagram of a wearable system

5. RF receiver – wireless signal receiver from the transmitter
6. RF working range is 1 km. The data can be transferred from one end to another in 1 km range.

4.2 Motion Detection Sensors

The motion detection sensors can be used for varieties of applications, including finding people's movement to Alarm siren, or switch on or off lights/fans, etc. The motion detection sensor is also used in the attire for the Visually Challenged to help them know the movements in front of them. More applications are Possible using this circuit and are left for those who can imagine well and include innovation (Figure 4.4).

1. PIR sensor – Passive infrared sensor
2. Arduino UNO –open-source microcontroller for combining our hardware and software
3. Vibrator – For notification
4. Battery – lithium-ion battery 3.7v rechargeable battery

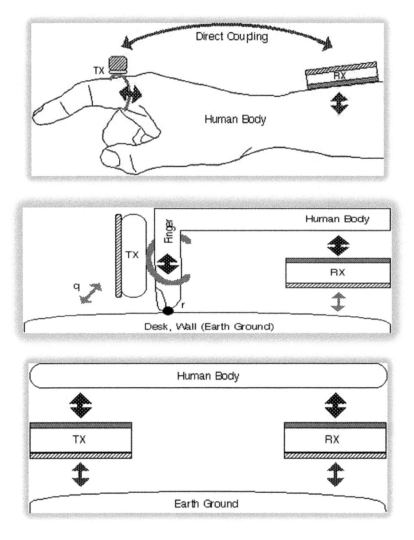

FIGURE 4.2
Functioning of wearable ring (Transmit TX/Receive RX)

4.3 Energy Saver/Rain Protection (Multiutility)

Similar to the motion sensors, this circuit helps to trigger activities based on events. Automatic viper blades and scrolling sticks in and out for the clothes-drying facilities in apartments are the conventional uses, but if one thinks it

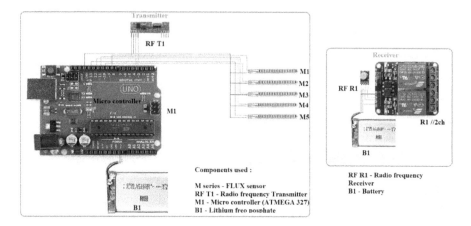

FIGURE 4.3
Circuit for flex sensor-based hand glove

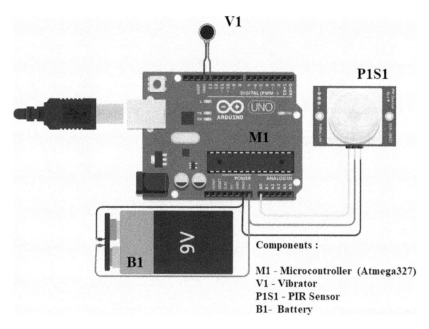

Components :

M1 - Microcontroller (Atmega327)
V1 - Vibrator
P1S1 - PIR Sensor
B1- Battery

FIGURE 4.4
Circuit for motion detection attire

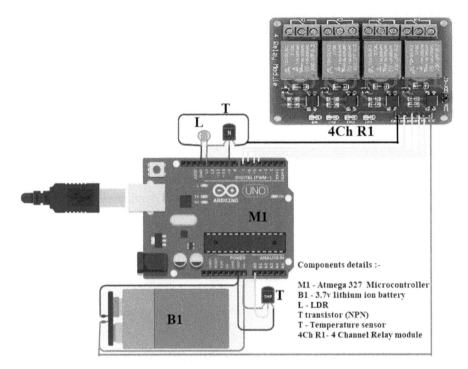

FIGURE 4.5
Circuit for energy saving/rain-triggered operations etc.

for wearable, there is a list of applications waiting to be developed. Figure 4.5 provides the view of the circuit.

L – Light-dependent resistor (LDR) (Automatic light on-off)

T – NPN transistor is used for rainwater automatic detection

T – Temperature measurement

M1 – microcontroller

B1 – Rechargeable battery 12v

4ch R1 – 4ch relay module – (automatic night light, Umbrella open–close, temperature low–high measurement) output load connection

4.4 Wearable Band

The band is a more sophisticated version of a hand glove and it works as a tracker for majority of applications. But the band is not limited to only the tracker but many safety applications could be developed with it. Figure 4.6 gives the basic idea about how it looks.

A1 &G1 –Mikroe MPU IMU click - MPU-6000 6-Axis Motion Sensor w/3-Axis Gyroscope and Accelerometer (It's used for measuring the values in hand)

M1 &D1 – Integrated microcontroller and OLED display

B1 – Buzzer for indication of values

B1 – 3.7v Rechargeable lithium-ion battery

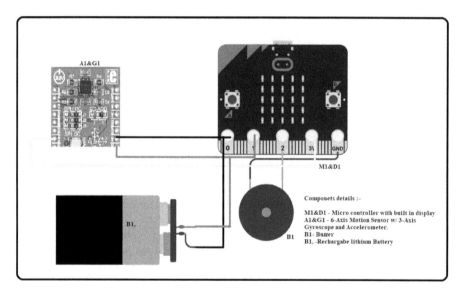

FIGURE 4.6
Circuit for multiutility wrist band

5

Materials and Components (Constructing a Simple Wearable Technology Device)

The basic components to make a wearable device are the shell, circuit, interface, and components. Mostly the shell, which is the device, holds all other elements of the wearable and is made up of metal, leather, or cloth. There are numerous sensors available nowadays in various sizes and shapes enabling us to design and develop innovative and a wide variety of wearables to suit any application.

In this section, some sensors and their uses are explained for a better understanding and to know how best they can be used in the wearables. Recent developments of carbon nanotubes play a better alternative to metallic sensors and thus make the wearables more thin, small, and comfortable to wear than their previous generation devices. Commonly sensors are made using very simple materials, including paper. The other materials in sensor making are polyethylene, polydimethylsiloxane, polyethylene naphthalate, polyethylene terephthalate (material used in making water bottles called PET bottles), polyimide, polyvinylidene fluoride, and polypyrrole (Figure 5.1).

Apart from the knowledge in electronics, the person who have a strong desire in developing wearables should be well versed and have the basic knowledge of other supporting materials, such as jumper wires, LEDs, circuit board varieties, power requirements, voltage analysis, resistors, and to use DMMs.

5.1 Flex/Bend Sensors

Flex sensor works on the changes in the resistance it can produce when it is bent since it consists of carbon-resistive elements. Flex sensor is a two-terminal device and does not have polarized terminals like diodes. So, there is no positive and negative which is an advantage in making them a part of wearables. Usually, P1 is connected to power source and P2 to ground. These flex sensors are usually available in two sizes. The first is 2.2 inch and another is 4.5 inch. There is no difference in the function of these two types of sensors. There are different varieties based on resistance value too. They are

DOI: 10.1201/9781003052906-5

(a)

(b)

FM Receiver & Finger Typing Detector

| | RF-BPF | PLL-RX | AF-BPF (90Hz) | Compa -rator |

PreAmp

Electrode (Signal side)

X5 (for each finger)

Electrode (Return side)

Symbol Generator (8bit CPU)

Symbol Output (Wireless Modem)

FIGURE 5.1
Presentation of a simple construction of wearable devices by student researchers

FIGURE 5.2
Pinout of flex sensor

categorized as LOW-resistance, MEDIUM-resistance, and HIGH-resistance sensors. These varieties are available to suit any appropriate application depending on the requirement (Figure 5.2).

5.2 Accelerometer/Magnetometer/Gyroscope

The smart phones we use change the display direction into portrait and land-scape according to the way we hold them. This smartness is brought into the phones by a small device called accelerometer. We are using accelerometers in the wearables to measure the acceleration of the wearables in different directions. As per Newton's law F= ma = Fs (on a spring when a mass is attached to it). Basically, acceleration is a function of displacement (rate of change of Displacement is velocity and rate of change of velocity is called Acceleration). The techniques to measure the displacement are resistive technique, capacitive technique, (measurement is done as the distance between the static and movable plates of a capacitor), and inductive technique. But nowadays the MEMS technique (which is a combination of micromechanical components attached to electronics) is used inside the accelerometer ICs. So, it is easy to measure the changes in capacitance and calculate the acceleration. Keeping multiple ICs in X, Y, and Z directions, we can easily identify the acceleration along with the direction.

There are two types of magnetometers, namely scalar and vector, both having a wide range of applications in measuring the magnetic fields with their sampling rate or cycle time. A significant use of magnetometers is in locating objects such as submarines, sunken ships, hazards for tunnel boring machines used for mining operations, road lying inside the mountain, or underground railway. It is also used in finding hazards in mineral and coal mines, unexploded landmines, unexploded artillery shells, live ordnance, and chemical toxic waste drums. We can find the presence of a wide range of mineral deposits and other geological structures. Magnetometers are devices used in devices like heart beat monitors, rifle, guns, and other applications like weapon systems positioning, automobiles using sensors in anti-locking brakes, and even in weather prediction (via solar cycles). They are present in steel pylons, drills used in oil, mineral, or water exploration to guide the drilling directions, archaeological excavations and searching, plate tectonics and radio wave propagation, and planetary exploration. The laboratory magnetometers help to determine the correct magnetic dipole moment of a sample with magnetic properties, typically as a function of temperature, magnetic field, or other parameters. This is done to reveal its magnetic properties such as ferromagnetism, anti-ferromagnetism, superconductivity, or other properties that affect magnetism.

Magnetometers are deployed in spacecraft for different applications, aircrafts like aeroplanes (*fixed wing* magnetometers) and helicopters (*stinger* and *bird*). Even they are used on the ground (*backpack*), towed at a distance behind quad bikes (ATVs) on a *sled* or *trailer*, lowered into boreholes (*tool, probe*, or *sonde*) and towed behind boats (*tow fish*).

Gyroscopes work more similar to the tops which young boys used to spin with a rope. The gyroscopes are used to maintain their orientation while they are spinning and can be used to measure the movements of the associated devices. Modern aircrafts and UAVs are making use of gyroscopes in their altitude indicators, heading indicators, and turn coordinators to maintain their orientations in comparing their orientation with the gyroscopes fitted inside them. But the gyroscope can also be used inside mini devices as small as wrist watches for maintaining precession. Gyroscope's applications are limitless, and it is used to maintain the balance of every object to which it is fixed. They protect human health starting from the balance of shoes to balance of exercise equipment to bring out symmetric results. They are even used in the assault rifles to maintain stability in their alignment of aim at the targets.

5.3 Vibration Sensors

Vibration sensors are used to monitor vibration in many different applications of the industrial process, power generation, marine, and in building service sectors. In every industry, the optimum performance of the machines has to be ensured for the best production. In order to maintain the smooth and desired performance, it is necessary to continuously monitor the parameters like speed, temperature, pressure, and vibration. Vibration is of two types one is axial vibration and the next is radial vibration. Vibration sensors use accelerometers, or eddy currents, or strain gauge to measure the vibrations. Apart from industry, the vibration sensors are used in construction of structures of roads and runways, automobile, clinical medicine, horticulture, aerospace engineering, etc. A proximity sensor strategically placed near the vibrating medium measures the eddy current by measuring the created high-frequency AC magnetic field and its variations (Figure 5.3).

The mechanical structure module as shown in Figure 5.4 can be replaced with an optical module in certain applications to measure the optical

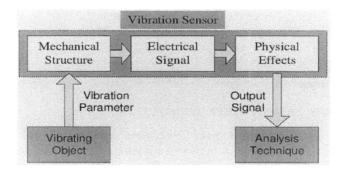

FIGURE 5.3
Measurement principle of vibration sensor
(Source: Himanshu Chaurasiya)

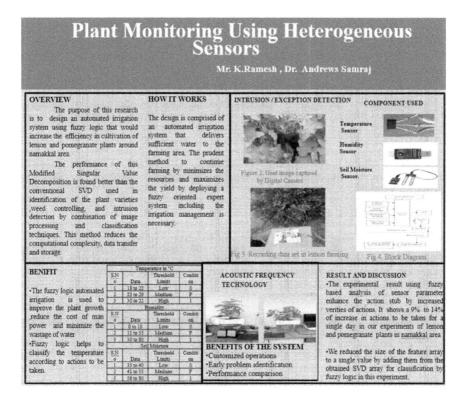

FIGURE 5.4
Application of Assorted sensors in agricultural Projects

quantities. A typical example for optical structure is the laser beam reflectors in the Doppler vibrometers.

5.4 Pressure Sensors/Force Sensors

A common use of pressure sensors is the transducers used in measuring the compressed air or gases and liquids in a container. The force sensor works on the principle that when a force is applied on an object with a specific mass, it changes the object's velocity. Other force parameters are thrust, drag, and Torque. While force is applied on an object, thrust increases the velocity of the object, whereas drag decreases the velocity and Torque generation changes in the rotational speed of the object. When there is a balanced distribution of forces in the object, no acceleration can be seen. The force sensors are designed using force-sensing resistors. These sensors consist of a sensing film and electrodes and convert the force into measurable quantities of resistance.

Pressure sensors can be made as simple as DIY projects at home using copper strips (tapes), jumper wires, a piece of velostat, and a think cord board. Sticking the copper tapes on a cord board and fold it to face each other with a gap in between them and connecting each piece of wire to the copper tapes at a distance from each other will form the base of the pressure sensor. A velostat should be sticked on the basic setup and sealed to make a perfect pressure sensor. Now the pressure applied on the setup can be measured using the wires connected to the circuit.

5.5 Smoke/Fire Sensors

These sensors include a simple heat sensor and variants to be built on the special wearable dresses of the firefighters and on many room surfaces where the alarm or protection against fire is required. We call them Flame Sensors, Smoke Sensors, Fire Alarms, etc., based on their working principle. They form the integrated safety equipment that help us in keeping our homes, offices, and stores safe from fire accidents. Some sensors work on the UV detection of radiation and are very sharp in detection sensitivity. A great natural quality makes the UV detector blind to sun's heat radiation but very sharp in identifying any other heat source radiation.

High-quality sensors like Honeywell 5800COMBO are used indoors and outdoors in sensing the narrow band radiation of heat sources. The advantages of the UV detector are its response to hydrocarbon, hydrogen, and metal fires, a high speed response time under 10 milliseconds, and blindness to solar radiation except when it is made to cover extra wide range. But there are some disadvantages also. The UV detectors trigger its response to welding even at a long range. Other responses include the lightning, X-rays, sparks, arcs, and corona. Even some gases and vapors will inhibit detection.

Other type of fire sensors are UV/IR detectors. They consist of a UV and single-frequency IR sensor paired together. The two sensors individually operate in the same way as previously described, but additional circuitry processes trigger from both sensors. This is to avoid the false alarm rejection capabilities than the individual UV or IR detectors.

Other ranges of fire detectors with much more combinations are also available. Even a visible fire detector with fire image processing technology is developed. A square law is applied when the distance between the fire and the fire detector is bigger compared to the fire dimension. Simple fire alarm circuits with Arduino UNO boards and free software are now available and effective.

5.6 Moisture Sensors

A Soil Moisture Sensor works on the principle of using capacitance in measuring dielectric permittivity of the surrounding on which the sensor electrodes are deployed. The presence of water molecules in the soil or any other medium can be found as a dielectric permittivity function. A proportional voltage will be created against the dielectric permittivity, and it represents the water presence of the medium. Many agricultural projects and plant monitoring works use moisture sensors to measure various parameters related to soil and plants, including water stress of plants.

The wholesale rate of a moisture sensor is as low as INR 25 in the market. A slight modification of the sensor and circuit makes it to work as a sweat (perspiration) sensor to be used in the wearable applications related to human body. Both clinical and augmentative (soft cyborg) applications use the sweat sensors. There are flexible forms of the sweat sensors available to form an appropriate component of the wearable.

Apart from sensors, many times, ICs, memory components, and batteries are incorporated as part of the wearables.

5.7 Design Considerations

Whatever may be the application or components used, it is the duty of the person who designs the wearable to ensure the complete function of the components and circuit boards used. It is better not to use any board or component if the entire operation or purpose for which it is being added to the circuit is not known.

A basic knowledge about the battery consumption, its sufficiency for a fair use, required recharging options, warning of misuse, fast draining warning, and recharging options are required in design process and such warnings need to be incorporated in the device design.

Finally, the possibility of misuse of the circuit design and application has to be widely researched, and all possible combat precautions have to be taken to avoid any misuse and failure of the purpose.

6

Assistive Technology and WT

Especially, the problem of communication with machines arises frequently in any human machine deployment, including health care and caregiving. As an example the problem in clinical health care deals with the communication of caregivers–support machines–patients. The problem grows enormous when the caregivers are allotted newly to a particular patient. The reason for this problem is the lack of communication that always takes place between the caregiver and the disabled/ elderly patients. In such communication, words become minimal, weak, and puzzled. They often use gestures to communicate when in pain or sick. These gestures are customized and have peculiar meaning which can be translucent only to the concerned pair of caregiver and the patient. It is apparent that the caregiver is bound to own responsibility of caretaking and can't transfer that easily to any third party. The experience in caregiving makes them experts indirectly. The problem becomes tougher when the expertise caregiving requires alterations or replacement. A skilled caregiver can be replaced only by a similar expert, which is hard to find under standard economic conditions. Severe errors may occur when the communication is not clearly understood or misinterpreted. The next degree of the problem worsening happens when the elderly patient is mute or has other special needs. So it is evident that the communication plays a vital role in caretaking process.

6.1 Wearable Electrodes for BCI/BMI

Brain–Machine Interfaces (BMIs) are the applications of wearables, coming under the cognitive ergonomics. The devices of BMI are designed with the intention of helping people who are disabled in nature. This BMI device enables disabled people to communicate with a computer or machine using their brains' electrical activities as the only medium. The technological solutions developed toward the rehabilitation engineering by combining the Clinical Neurophysiology with computer applications for the problems faced by individuals with disabilities lead to these Brain–Computer Interfaces (BCI). Clinical practices frequently involve Electroencephalogram (EEG),

DOI: 10.1201/9781003052906-6

which is the electrical impulse of the brain and can be recorded by electrodes placed on the scalp. A typical labeled position of the electrodes is shown in Figure 6.1.

Initially the EEG signals were generally investigated for the diagnosis of mental conditions such as epilepsy, memory impairments and sleep disorders. In 1929 Hans Berger was the first person who recorded the (EEG in humans. The trend of research using EEG is the development of the assistive technology and the rehabilitation application called Brain–Computer Interface (BCI sometime called BMI) started by researchers in the late '90s. Birbaumer, in 1999, and Vaughan and Wolpaw, in 2006, were the leading scientists in this area. BCIs are constructed by recognizing the electrical, magnetic, and physical manifestations of the brain activity.

It uses the time, frequency, and spatial features of the brain. The BCI has been defined as the exceptional communication system that does not require any conventional muscle activity (www.bciresearch.org, 2008).

EEG-based BCI designs are aimed at people who require a basic communication link, which they are deprived of and a hands-off device control as they use the electrical activity of the brain to interface with the external environment, therefore circumventing the use of peripheral muscles and limbs.

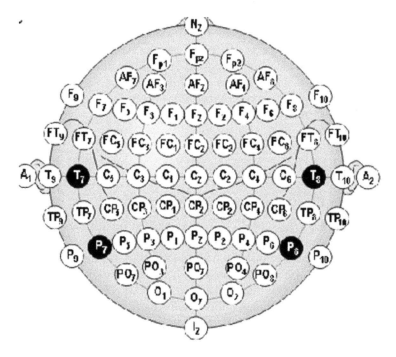

FIGURE 6.1
64 Sensor Electrode positions

There are BCIs developed and tested on animals like monkeys and rats too by scientist Serruya, M.D. in 2002, D.M. Santucci in 2005, Gopal santhanam in 2006, and T.C. Marzullo in 2005. Many current applications of BCIs are in communication system and are for locked-in individuals. These paralyzed people who are forbidden to communicate with their surroundings can make use of the techniques like speller paradigms to communicate and perform device control such as wheelchair movement, prosthetics control, and other rehabilitation engineering & assistive technology devices. Generally, for the common community, some of the potential applications of such bionics are hands-off menu selection, flight/space craft control, cyber security and virtual reality entertainment as identified by Willingham, D.B. in 2001. Palaniappan and Mandic were even used BCI in biometrics applications in 2007.

The BCIs used in the field of assistive technology or rehabilitation engineering extremely depend on wearable technology both in the case of invasive and non-invasive methods. Generally, the non-invasive techniques are in practice nowadays with a more sensitized equipment that the BCIs will function to assist the neurological patients in day-to-day life. In recent advancements, there has been a great demand and dynamic growth of interest in producing revolutionarily innovative devices which would enable people to control computers (and therefore any other reactive machinery) by their brain activity alone. This benefits the severely disabled and even common

FIGURE 6.2
Rehabilitation experiments through BCI

people to control devices based on their brain activity rather than depending on the limbs (Figure 6.2).

Various techniques are available for capturing brain activities, which include EEG, functional Magnetic Resonance Imaging (fMRI), Magnetoencephalography (MEG), Positron Emission Tomography (PET), and Electroocculogram (EoG). Among these techniques, EEG is the most preferred for BCI designs, because of its non-invasiveness, cost-effectiveness, and easy implementation. The non-invasive BCI methods use EEG, MEG, PET, fMRI, and optimal imaging (near-infrared spectroscopy (NIRS)). Invasive BCI is based on electrocorticogram (ECoG), though effective due to health hazards as the electrodes are surgically implanted, this method is used only for crucial clinical needs only.

This is one of the reasons which encourage many to concentrate on the design of an EEG-based BCI for applications in assistive technology.

A BCI's design is usually realized by using Visual Evoked Potentials (VEPs) or movement-related potentials (MRPs). The clear representations of MRP can be observed in the EEG's μ -rhythm (8–12Hz) and/or Beta rhythm when a person performs a motor activity or imagines a motor activity. Such an activity can be easily captured from the EEG's channels C3 and C4. However, the VEP could be realized in the Pz, Cz, Fz channels and their surrounded positions. Here comes the need of an efficient wearable technology, where these electrodes are to be made in contact with the exact locations of brain by means of a wearable electrode cap.

Current designs of BCI totally depend on the wearable technology and have four main stages, which are raw signal acquisition through the wearable electrodes, signal preprocessing or conditioning, feature extraction, and classification of features into intended activity or communication terms. Among these stages, feature extraction and classification methods play an important role because, a successful BCI depends on its ability to extract accurate EEG features according to different tasks and to efficiently classify them in a real-time environment.

One of the most challenging tasks in designing a BCI is in choosing the relevant features from the EEG signals, which are chaotic in nature. Accurate feature extraction (AAR) method is very important in determining the performance of a classifier. Incorrect features could lead the classifier to have poor generalization, computational complexity, and requires a large number of training dataset to achieve a given accuracy (Figure 6.3).

FIGURE 6.3
A sample BCI design using AAR for feature extraction

The wearable electrode cap in BCI is found necessary for the following reasons:

1. The BCI requires continuous input of brain signals.
2. The electrodes need to have a socket to stick on to the scalp.
3. The electrodes have to be in their exact position as the brain activity depends equally on spatial too.
4. Those who need such BCI that are not able to use the electrodes by their own and some assistance is required for them to position them on their head.

6.1.1 EEG Rhythms

The frequency ranges of EEG are divided into different bands, namely the Delta band (0.1Hz-3Hz-5Hz) that usually occurs during sleeping; Theta band (4–7.5Hz), when the subject is almost unconscious; Alpha band (8–13Hz), known as internally focused state; Beta band (14–30Hz) occurs when active thought, alertness, and visual scanning is in progress; and Gamma band (>30Hz), for example, when the focus for motor imaginary BCI is on the specific part of the Alpha band (10–12Hz) which is the μ (Mu)-rhythm. This μ-rhythm can be recorded over the sensorimotor cortex (electrodes C3 and C4). The main advantage of the mu-rhythm over other rhythms is that it does not appear to be influenced by eye-blinking and users can learn to voluntarily control the rhythms after receiving training to some extent. This was reported by Pineda J.A. in 2003.

There are several methodologies for implementing EEG-based BCI. They are, evoked potentials (typically from visual stimulus, auditory stimulus sometimes), better known as VEP used by Donchin and others in 2000, Wangand and others in 2006, mental activity by Palaniappan in 2006, motor imagery by Wolpawand and coresearchers in 2002, Kubler, A. and others in 2005, R. Sitaram and coresearchers in 2007, and Slow Cortical Potentials used by Mensh in 2002. The VEP approach mentioned above could be further divided into P300-based VEP used by Donchin and coresearchers from year 2000 and steady state VEP (SSVEP) used by Wang and coresearchers in 2006.

6.1.2 Signal Acquisition by Wearable Methods

The EEG is an electrical activity recording from brain usually through multiple numbers of electrodes placed in different locations of brain either by invasive or non-invasive way. All the capacity and disorders of brain will be reflected in the EEG and that's why it is useful in diagnosis of problems like stroke, head injuries, brain tumor, brain damage due to head injury, brain dysfunction due to many clinical reasons, encephalitis (inflammation in brain), and sleep disorders. EEG is the safest method and causes no side effects or pain in the patients who undergo the recordings.

Subjects will generate brain activity through an experimental paradigm that would depend on the particular BCI approach. The protocol to be followed by the subjects could be thinking about making imaginary movements, focusing on flashing characters on a screen, etc. This brain activity will be picked by electrodes (normally Ag/AgCl and nowadays Au) placed on the scalp. The placement of electrodes commonly follows the 10–20 system (19 active electrodes) or extensions of this system (32, 64, 128, or 256 active electrodes). The recordings are normally referenced to the left and/or right mastoids. An example of the 10–20 electrode placement system is shown in Figures 6.4 and 6.5.

An evoked potential, sometimes called an evoked response is a small electrical potential created as a specific pattern of response from a particular location of the nervous system, mostly from the brain of humans or birds or animals due to the presented stimulus such as a visual flash of light or a sound with a sinusoidal waveform called pure tone. The stimulus is a detectable change felt in the physical or chemical structure of an organism's internal or external surroundings. The brain is capable of responding by different types of potentials resulting from stimuli of different modalities and types presented to it. Among such potentials, the Evoked Potential (EP) is very distinct from other spontaneous potentials. Usually, these EPs are detected through the recordings of electroencephalography (EEG, ECoG, electromyography (EMG), or other electrophysiologic methods.

Such evoked potentials are used for electrodiagnosis and keen monitoring for detecting the disease and drug-related sensory dysfunction and intraoperative monitoring of sensory pathway integrity.

All these recordings are done with the help of a wearable electrode cap that is fitted with 16/32/64 electrodes based on their position and necessity. Each electrode is named distinctly after its position (Figure 6.6).

6.2 BCI – Example 1: Single Trial Source Separations of Visual Evoked Potential Signals Using Various Methods and Techniques to Identify Controls from Alcoholics

Single trial source separation problem of VEP signals has been addressed as Principal Component Analysis (PCA). Here, in Section 6.2, as the first example, a BCI using VEP has been presented. This new method of BCI called spectral power ratio (SPR) selects and uses specific principal components (PCs) from the PCA process to perform the accurate source separation. This

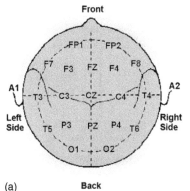

(a)

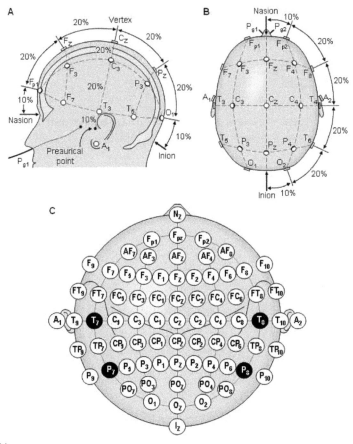

(b)

FIGURE 6.4
Electrode placement system and 75 electrode positions (*American Electroencephalographic Society.*)

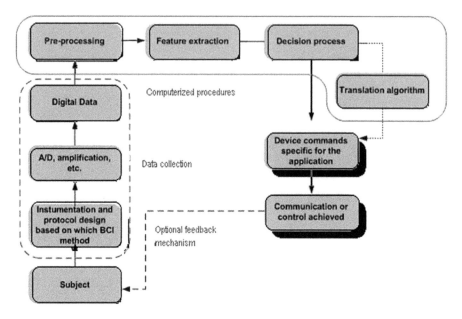

FIGURE 6.5
BCI – Example 1: Single trial source separations of Visual Evoked Potential (VEP) signals using various methods and techniques (a BCI wearable Application to identify controls from alcoholic)

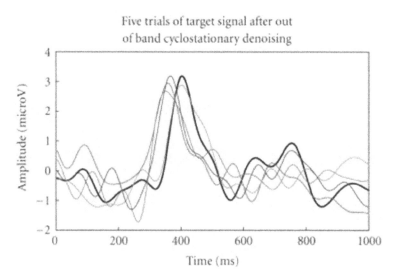

FIGURE 6.6
Overlaps of five trials of P300 evoked EEG signals

SPR method is applied on the EEG signals in which the VEP is buried. A preliminary investigation is performed with the new method to test its capability to separate artificial VEP signals contaminated with background EEG signals. The focus was on extracting the P3 parameters. The new SPR method suggested resulted in increased Signal-to-Noise Ratio (SNR) as compared to other existing methods like Kaiser (KSR) and Residual Power (RP) for selecting PCs. The SPR's performance is better especially when the noise (i.e. EEGs) is high. With this confirmation, the SPR method is applied to real VEP signals naturally buried in EEG. The SPR technique's potential to identify the P300 peak from the real VEP signals to analyze the latency responses for a paradigm of matched and non-matched stimuli by pictures. The P3 parameters extracted through the SPR method showed faster and higher P3 response for matched stimuli, which confirms to the existing neuroscience knowledge. So, it is confirmed that the use of SPR technique of PCA to use in single trial source separation of EEG signals and a better BCI can be developed than with the PCA using KSR and RP methods since they did not indicate any difference for the stimuli.

In the early days of BCI research and development, two major methodologies were followed. One is ensemble averaging and another is the single trial analysis. The averaging method follows a few trials of the same brain activity and the analysis average is considered as the final decision. The very reason for following the averaging is to avoid human errors and for the confirmation of the intended brain activity to ensure the decision making. On the other hand, the single trial analysis is to reduce the time of recognition of the intension of the subject operating the BCI. This is to make the BCI faster and reduce boredom of the subjects.

The risks involved in single trial are the mistakes that may happen in decision making due to similar responses received from the subject and there may be dilemma in concluding it on a choice from more than one target.

As mentioned above, the basic method of solving the contamination of EEG in VEPs was through averaging. However, averaging requires many trials and it might distort the single trial information.

Many researchers adopted methods using PCA for a single trial analysis of VEP separation from EEG. This method was found suitable by the researchers since it resulted in a good single trial analysis of VEP signals.

In this demonstration, an advanced step further in single trial analysis by using a PCA variant for performing single trial analysis of VEP signals was introduced. The existing popular procedures to select PCs in PCA are Kaiser (KSR) and Residual Power (RP). Here, in this presented example, a new variant of PCA called Spectral Power Ratio (SPR) was used during the single trial experiments; the results show that the KSR and RP methods are unable to retain their performance when the content of background EEGs is high.

The very reason for this work is to find a more efficient method of selecting PCs for the best possible reconstruction of the VEP as during the EEG noise is high.

This SPR method chooses the proper PCs for the effectual rebuilding of the source VEP signal when the expected noise contamination level goes beyond the capacity which Ksr or Rp methods could handle. This SPR technique aids in extracting undistorted VEP feature by choosing selective PCs from the group of PCs calculated by PCA. This is of great help in establishing a single trial model in neuropsychological and clinical applications dealing with VEP.

6.2.1 The Methods Involved in This BCI Demonstration

As mentioned before, the entire demonstration starts with a set of artificial VEP signals created and mixed with the real EEG noise. This is to prove the effectiveness of the newly adopted SPR method through a VEP extraction simulation study. SNR calculation is easy since the artificial signals are created before the experiment and are compared with the extracted VEP feature by the SPR. This SNR value is used to compare the advantage of the adopted SPR technique with KSR and RP methods in selecting PCs. Since the fitness of SPR is confirmed by the experiments, as a next step, the SPR method is applied on signals that contain real VEP components with P3 amplitude and latency responses buried in EEG noise, obtained during a matched and non-matched stimuli experiment.

6.2.2 Artificial VEP Simulation

Sixty-four artificial VEP signals were created using different combinations of Gaussian waveforms, each with different mean, variance, and amplitude. These basic waveforms were created using the equation

$$G(n) = \left(A / sqrt\left(2\pi\sigma^2\right)\right)\exp\left(-\left((n-\mu)^2\right)/2\sigma^2\right) \qquad (6.1)$$

These VEPs were limited to 8 Hz to simulate P3 responses, which are limited to 8 Hz.

These artificial VEPs were mixed with the real EEG signals, which were obtained when the subjects were at rest to avoid the presence of any similar potentials in it. These EEG signals were whitened to remove their correlation, before adding to the artificial VEP signals,

$$W(n)_{VEP+EEG} = X(n)_{VEP} + Y(n)_{EEG} \qquad (6.2)$$

Then the contaminated signal, W, was normalized to zero mean and unit variance.

$$W = (W - mean(W))/Std(W) \tag{6.3}$$

6.2.2.1 Applying Principal Component Analysis

PCA to extract VEP signals from EEGs was carried out. First, the covariance of the signal W was computed using

$$R = E(WW^T) \tag{6.4}$$

Let F be the orthogonal matrix of eigen vectors of R and D and the diagonal matrix of its eigenvalues $D= diag\ (d_1....d_n)$. Then the PCs could be computed using

$$Y = F^T W^T \tag{6.5}$$

Some of the PCs will represent the VEP and some will represent the EEG. The selection of required PCs from all available PCs was done by three methods: KSR, RP, and SPR.

These selected PCs were then used in reconstruction, where the reconstructed signal now ideally contains only VEP. The reconstruction was done using

$$X = FF^T YY^T \tag{6.6}$$

where the FF and YY corresponds to the selected eigenvectors and PCs. How PCs are selected for reconstruction

(i) **Percentage of total residual power retained (RP)**
In the RP method, the first few sorted PCs were selected where the percentage of eigen values selected covers the 95% of the overall eigen values of the PCs. The remaining PCs with very small eigen values were omitted and only the selected PCs were used for the reconstruction of the VEP signals.

(ii) **Kaiser's rule (KSR)**
Using the KSR method, the PCs selection was done such that the eigen values of selected PCs were not less than 1.0. The PCs with less value were omitted and only the selected PCs with eigen value more than 1.0 were used for the reconstruction of VEP signals.

(iii) **Spectral Power Ratio (SPR)**
In SPR method, irrespective of the eigen value, only the PCs that contained significant amount of 0–8 Hz spectral power were selected. This

frequency limit could be varied according to the purpose. In this demonstration, since the P300 components are considered, the limit was set to 8Hz. After some experimental simulations, it was found that the values 0.5–0.6 were sufficient as thresholds for finding the 0–8Hz spectral power, i.e. for selecting the PC under consideration, if the ratio of spectral power below 8 Hz over the total spectral power exceeded this threshold, then that PC would be selected.

The other PCs with SPR below this threshold were reset to zero. As a next step, these selected PCs were used to reconstruct the VEP signals.

6.2.3 Signal-to-Noise Ratio Calculation

In order to compute the efficiency of the three different PC selection methods, SNR computations were carried out for the reconstructed VEP signals. This calculation was implemented by

$$SNR = 10\log_{10}\left(Variance(X)/Variance(W-X)\right) \tag{6.7}$$

The total SNR for all the 64 reconstructed VEP signals were also calculated for each method.

Introducing Different Noise Factors

The entire experiment was repeated with the signal W but adding noisier EEG signals, i.e. EEG signals with amplitude multiples of 2, 5, and 10:

$$W(n)_{VEP+noise} = X(n)_{VEP} + NY(n)_{noise} \tag{6.8}$$

where $N = 2, 5,$ and 10.

Again, the performances of all the three methods were obtained for comparisons. The experiments with artificial VEP signals were over here and the fitness of SPR to apply it on real VEP signals buried in EEG noise is confirmed.

6.2.4 Single Trial P3 Responses Experiment Using Real VEP

The experiment is again repeated using all the three PC selection methods on the real VEP signals which were recorded from different subjects while being exposed to two types of stimuli. The two types of stimuli were by using pictures of objects chosen from Snodgrass and Vanderwart picture set for display. Figure 6.7 shows some of these pictures.

The experiment that is conducted to collect the VEPs from different subjects is a matching and non-matching picture showing session. During this session, using a wearable electrode cap the EEG signal is collected from multiple channels in which the VEP may present based on the visual stimuli given to the subjects through the pictures. The first visual stimulus (S1) shown to the subjects were a few randomly chosen pictures. The second stimulus shown was chosen from both the matching (S2M) and the non-matching (S2N) rule relative to the initial stimulus (S1). The second stimulus picture set may contain the same pictures shown in the first few pictures (S1). When the same picture from (S1) is shown in the second stimulus test, it is known as (S2M). If a different picture is shown in the second stimulus which is not shown in the few pictures during S1, it is known as (S2N). To reduce the possible ambiguity, S2N was chosen to be different from S1 not only in its visual appearance but also in terms of the semantics. For example, if a picture of an elephant is shown for S1, then S2N will not be a picture from the animal category. One-second measurements of EEG signals from the wearable electrode cap after each stimulus presentation were recorded. This experiment was done with randomly selected few trials from four different subjects. S2N stimuli are explained in Figure 6.7.

The eye-blink artifact contaminated VEP signals were removed from the records, since these EoG are not necessary for this application and also contaminate the VEP. These signals were detected based on amplitude discrimination (the threshold value of 100 µV was used for this since blinking typically produces potential of 100–200 µV lasting for 250 ms). In this juncture, it is necessary to remember that there are other wearable applications possible with such EoG.

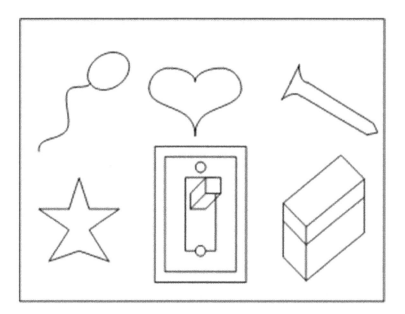

FIGURE 6.7
Some pictures from Snodgrass and Vander wart picture set

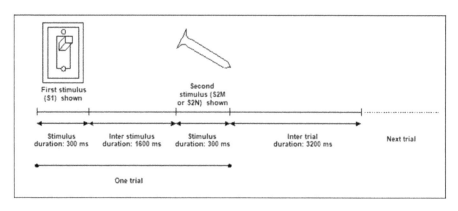

FIGURE 6.8
Example of stimulus presentation for the case of S2N

The normalization of data is essential to do the processing into a scale, this could be done by setting the pre-stimulus baseline to zero; the data were made zero mean. Following the approach of the researchers like Begleiter, where P3 responses were shown to be band-limited to 8 Hz, the extracted VEP signals from S2M and S2N stimuli were low pass filtered using a combination of a 9th order forward and 9th order reverse Butter worth digital filter with a cutoff frequency at 8 Hz. This way, a minimum attenuation of 30 dB was achieved in the stop band, with the transition band being between 8 and 12 Hz. Notice that the reason for both forward and reverse filtering was performed to ensure no phase distortion.

Though the experiment is done using a wearable electrode cap having multiple electrodes to record EEG, single trials of VEPs from the Pz channel only were analyzed in this experiment, because the P3 response reaches its maximum in the midline parietal area and Pz is the electrode positioned there to record it and to make the experiment simple. The amplitude and latency of P3 responses were detected via an automated procedure, whereby this P3 component was identified as the largest positive peak in the period of 300–600 ms after the stimulus onset. A t-test was conducted on the observed amplitude and latency readings to establish a statistical difference in P3 amplitudes and latencies between stimuli S2M and S2N. Figure 6.8 shows an example of the experimental details of stimuli presentation.

6.2.5 Outcome of the BCI Experiment 1

The performance results of the experiments done with artificial VEP analysis are given in Tables 6.1, 6.2, 6.3, 6.4 and shown in Figures 6.9, 6.10, 6.11, 6.12, 6.13, 6.14. The number of VEP signals is restricted to 4 as an explanation

TABLE 6.1

Comparison of SNRs of RP, KSR, and SPR methods with EEG factor=1

	SNR			
Randomly selected channels	Original	*RP*	*KSR*	*SPR*
1	0	7.67	12.69	12.69
2	0	2.92	9.16	9.16
3	0	2.93	12.57	12.57
4	0	1.98	10.60	10.60
Total (64 signals)	**0**	**185**	**698.71**	**698.71**
Average (64 signals)	**0**	**2.89**	**10.91**	**10.91**

TABLE 6.2

Comparison of SNRs of RP, KSR, and SPR methods with EEG factor=2

	SNR			
Randomly selected channels	Original	*RP*	*KSR*	*SPR*
1	−6.02	2.59	8.32	8.76
2	−6.02	−0.07	6.35	6.48
3	−6.02	0.04	5.28	5.83
4	−6.02	0.32	4.83	5.21
Total (64 signals)	**−385.31**	**−13.43**	**390.45**	**428.32**
Average (64 signals)	**−6.02**	**−0.20**	**6.10**	**6.69**

TABLE 6.3

Comparison of SNRs of RP, KSR, and SPR methods with EEG factor=5

	SNR			
Randomly selected channels	Original	*RP*	*KSR*	*SPR*
1	−13.97	−1.24	3.98	4.01
2	−13.97	−1.54	3.79	4.16
3	−13.97	−1.66	3.14	4.47
4	−13.97	−1.58	2.53	3.16
Total (64 signals)	**−894.68**	**−120.67**	**61.64**	**107.17**
Average (64 signals)	**−13.97**	**−1.88**	**−0.9**	**1.67**

TABLE 6.4

Comparison of SNRs of RP, KSR, and SPR methods with EEG factor=10

	SNR			
Randomly selected channels	Original	RP	KSR	SPR
1	−20	−1.79	2.28	2.37
2	−20	−2.13	1.64	2.49
3	−20	−2.28	−0.23	1.08
4	−20	−2.50	−2.40	−0.85
Total (64 signals)	**−1280**	**−146.69**	**−51.91**	**−27.20**
Average (64 signals)	**−20**	**−2.29**	**−0.81**	**−0.42**

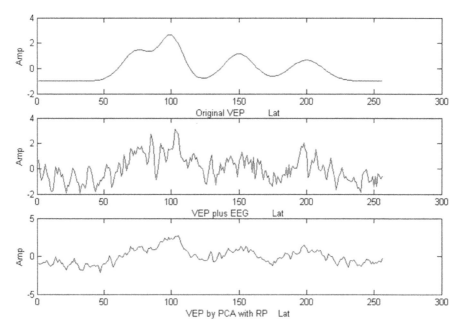

FIGURE 6.9

Artificial VEP signals with PCs selected using RP (EEG factor of 1)

sample and the total and average for 64 artificial VEP signals are provided. The evidence of the effectiveness of the adopted SPR method could be seen from the higher SNR values as compared to the original contaminated signal, KSR and RP methods.

It is also clear that the SPR method gives improved performance in comparison to KSR and RP, when the EEGs were higher.

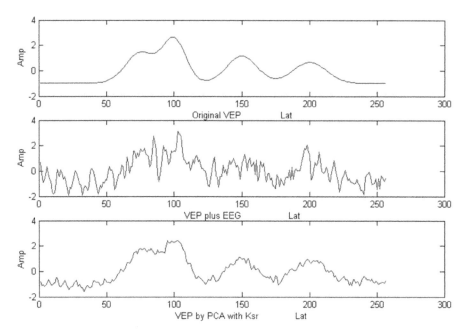

FIGURE 6.10
Artificial VEP signals with PCs selected using KSR (EEG factor of 1)

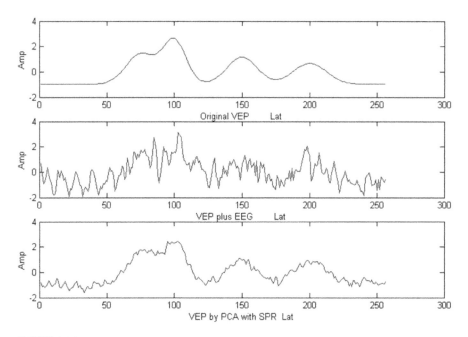

FIGURE 6.11
Artificial VEP signals with PCs selected using SPR (EEG factor of 1)

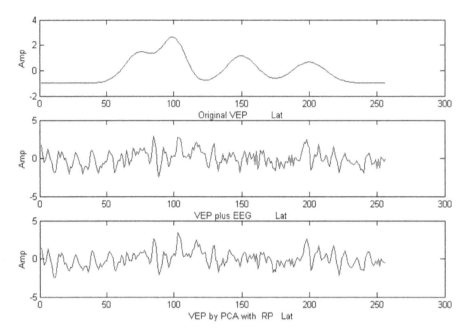

FIGURE 6.12
Artificial VEP signals with PCs selected using RP (EEG factor of 5)

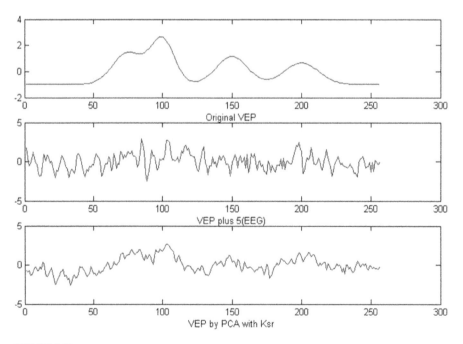

FIGURE 6.13
Artificial VEP signals with PCs selected using KSR (EEG factor of 5)

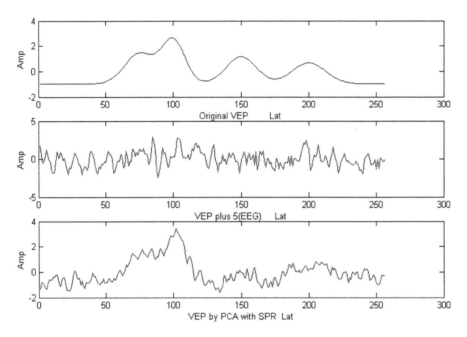

FIGURE 6.14
Artificial VEP signals with PCs selected using SPR (EEG factor of 5)

TABLE 6.5

T-test results of P3 latencies and amplitudes for stimuli S2M and S2N

Subject	RP		KSR		SPR	
	Latency	Amplitude	Latency	Amplitude	Latency	Amplitude
1	0.3397	0.8219	0.5111	0.7844	0.9978	0.0034
2	0.8798	0.3063	0.5980	0.4561	0.9994	1.89e-015
3	0.1425	0.1814	0.0546	0.8401	1.0000	8.11e-004
4	0.4590	0.4142	0.7279	0.5998	0.9683	1.26e-010

Table 6.5 gives the result for the t-test analysis of P3 latencies and amplitudes. The hypothesis tested for latency is S2N>S2M, while for amplitude, the hypothesis tested is S2M>S2N. The results indicate that only P3 parameters extracted using PCA from SPR method show difference, while KSR and RP methods do not indicate any differences. The results using SPR method show that P3 amplitude was higher from S2M as compared to S2N (with p<0.05). In addition, it also shows that P3 latencies are smaller (i.e. P3 responses are faster) for S2M as compared to S2N (with p>0.95). These results confirm to the other studies conducted with matched and non-matched stimuli.

Therefore, in conclusion, the SPR–PCA-based reconstruction of VEP signals is suitable for extracting single trials of VEP with its undistorted amplitude and latency as long as we know the frequency range of the parameter that we wish to extract.

6.3 BCI Example 2: Speller Paradigm

The second example of a BCI using the wearable electrode cap is an interesting communication demonstration experiment in which the people of different locked-in problems could communicate through an unconventional means called speller paradigm focus. Figure 6.15 shows an example of such speller and picture paradigms, which works on a flashing mechanism. The flashing mechanism works in such a way that the rows and columns are sequentially and periodically highlighted and other rows and columns go dim when a particular row or column is highlighted. Highlighting is a process of illuminating by making the column or row bright during the time interval when it is selected; for example, the row 1 is highlighted first for two seconds and after two seconds it goes dim along with the other rows and columns. The second row is selected immediately after the first row goes back to the dim state for illumination, when the next 2 seconds pass the third row gets a change to glow while row 2 goes to dim along with other rows. Once all the rows finish their chances of illumination, the time for columns starts, the same time limit is given for columns and each column gets illuminated one by one for two seconds. No column or row is illuminated at the same time.

This speller paradigm is used to help the users to invoke their Visually Evoked Potentials from brain while they gaze the particular letter present at the paradigm that is illuminated twice by the row and column of that desired letter or digit.

(a)   (b)

FIGURE 6.15
A typical example of a speller and picture paradigms for VEP invocation

6.4 BCI Example 3: Picture Paradigm

The third example of a BCI using the wearable electrode cap is a picture paradigm. Figure 6.15 b shows a typical example of a picture paradigm which many researchers found fruitful in their BCI experiments using VEP. The paradigm works in the same row by row and column by column illumination explained in Section 6.3.

A good explanation and working of both the speller and picture paradigms are explained in Section 6.5 for an application that creates a passcode for authentication by means of VEP-based BCI.

6.5 Example 4: Bio-Cyber Machine Gun – A New Mode of Authentication Access Using Visual Evoked Potentials

The Bio-Cyber Machine Gun (**BCMG**) is designed to work as a defensive tool used to protect misuse of authentication, Access Control, in aiding cryptography and information hiding by means of password shooting. This application is very similar to the previous two examples but differs only in interface and application. The need for this tool amidst various existing authentication protection methods is unique. The conventional password, pin number, smart card, barcodes or biometric fingerprints, palm print, iris pattern, and face recognition are only used for a specific purpose of authentication or access control on a one at a time basis. The amounts of information we can pass through these methods are also very limited. The ever-growing demand for integration of services and higher-level security needs leads to the lookout for an efficient and reliable system, which can do multiple tasks in a highly secured way. This kind of sensitive and complex systems can be made effective, robust, but simple when we use them from a very close mode like biometrics of a person, but without the major drawbacks of biometrics. Moreover, this is the only possible authentication and access control method, which can be used by patients, paralyzed people, and aged disabled people who cannot adopt the conventional methods.

The background of this system development is an insisting demand for finding an alternative authentication for the people who cannot use the conventional way of logging into a system due to their disability. At the same time, normal humans are also capable of using this system so that the usability is universal. The human body generates many kinds of signals known as in general biosignals, including signals from Heart beat (ECG), and from Brain (EEG) and other few. Most of these biosignals are independent and automatic and cannot be composed to a fixed rhythm by humans. But some signals generated by muscle activities that can be under the control of a person can be made rhythmical. But the production methods are not secured and protected for example EoG can be simple and rhythmic but anyone can learn it from observation of the users eye blinks or movements. Therefore, BCMG uses VEP signals for the above purpose and found it to be feasible and appropriate for sensitive multipurpose security systems.

The VEP normally buried in EEG can be made rhythmic by using an oddball paradigm that gives a visual stimulus. It is faster than the mental prosthesis method. A conventional Light Machine Gun (LMG) can fire at the rate of 300–500 rounds per minute in its rapid-fire mode (R) using a belt supply. We use this model of rapid-fire mode to randomly activate the paradigm to evoke the VEP from brain to coin the current password at any moment.

The procedure and benefits of BCMG are as follows. The BCMG was designed for this purpose as a tool using two major components: one is the signal capturing unit and the second is an interface unit. Using the first unit, the raw EEG recordings are taken from the scalp. These signals always show some artifacts of signals around 100Hz, which may denote the eye blinks and will be eliminated by filters.

The electrodes fixed on the parietal area are used to record these EEG signals for one second immediately after the visual stimulus. Using the interface unit, these recordings are bandpass filtered to remove artifacts and translated after separating the VEP using the latest methods of feature extraction by variants of PCA into rhythmic control codes with respect to the speller paradigm, which are considered as the bullets of the BCMG. Whenever the signal peak at P300 goes high, the row or column of the spelling paradigm is considered for its characters and the intersecting character is selected as the target letter for coining the passcode. This paradigm is shown in Figure 6.15. In the testing phase, the translated and coined passcodes like 'CAT' are stored and labeled. The experiment is repeated to reproduce the codes that are compared with the first set of control codes available in the database using a matching and non-matching routine. The average accuracy of reproducing 40 codes having three to five characters each is 90.24%. The similar procedures are followed to capture signals by changing the character paradigm to a picture paradigm shown in Figure 6.15 b, and the accuracy of finding 10 pictures is 100%.

The first and third signals shown in Figures 6.16 and 6.17 are the non-target representation of the spelling paradigm and the second signal is the target representation. By observing Figure 6.17, it is understood that the second signal ends with a positive amplitude; it is impossible to recognize, fake, or reproduce such signals by a bare human eye understanding. Hence, these signals are considered as rhythmic bullets that can be emitted by the human brain in a rapid speed when there is a visual stimulus. These rhythmic bullets can be synchronized with any application for any individuals to use it as a guarding weapon to protect their electronic payments, equipment, systems, information, and files, etc.

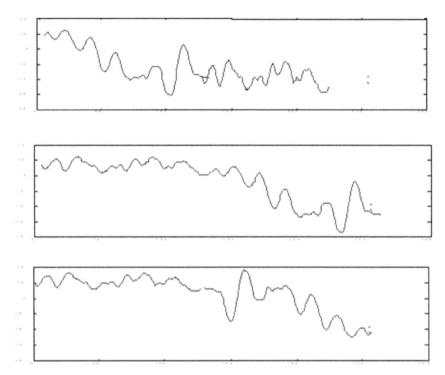

FIGURE 6.16
Brain signal responses with P300 response potentials and without potentials

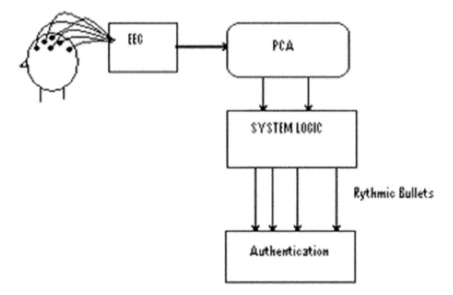

FIGURE 6.17
Setup for creating rhythmic bullets of VEP

The advantage of using this BCMG tool is its ability and efficiency in multitier password-protected environments. Basically, producing various combinations of these bullets involves no cost. The H/W device used to capture these signals is a few electrodes engraved wearable cap or helmet only or just a patch wearable on scalp. This may aid the protection in generating the signals since no one can see and recognize the signal generated. It is safer than generating signals using finger movements or any other muscle movements. It is easy and faster than generating biometric pass codes that need scanning.

Technical advantages are also more by using this bludgeon. Multiple levels of authentication are possible in this since we use different bullets to hit different target authentication, which will make it difficult for the hacker to access to the system easily. This is just like using different kinds of locks and keys for every door in the house. A wide variety of signals provide huge bandwidth of usage so no need to worry about limitations. These bullets can be changed from time to time for the same level of authentication so that we can arrive at a highly secured and reliable system at every time of use and is better than the cancelable biometrics since no need to use more than one factor for authentication. Another advantage here is that no need for a media to store and carry this ever-changing authentication. Hence, it is safer than any form of biometrics.

The future trends of this system will lead to a multitier combinational access code, which will serve as a multipurpose smart card for the aged and disabled people. These systems can be embedded on any kind of devices such as wheelchairs and will help the users to program their activities including their secured operations while they are in their bed well in advance before they start going out to perform their work.

In conclusion, this example BCI system will meet up with the arising requirement for multiple pass codes and pin numbers for access control and authentications in near future. It is very difficult to possess more numbers of such pins or codes on different or same media which is dangerous too. The BMCG will be the suitable, safe, and cost-effective solution for this purpose. It is not only safe and cheap but also user friendly and very much suitable for aged and disabled people who cannot handle the other authentication tools properly.

6.6 BCI Example 5a: Motor Imagery – Adaptive Bandpass Filter

BMIs are visualized as an unconventional, direct communication crossing point between a machine and the brain of an individual. A BCI/BMI that is always under consideration of automatic machine control is Motor Imagery. The left and right side of human brain are the spots where the control of limbs and their motor movements triggers out. A novel approach of a classification of Motor Imaginary Signals from brain for Machine Communication is given in this section.

A great interest for BMI that controls machines like vehicles and wheelchairs is rising nowadays, activated by multiple hopeful scientific and technological results. Many methods adopted in quantification, processing, and categorizing brain activities to understand the human intentions from the neuronal signals are in practice. Interpreting such brain signals into machine control is always regarded as a challenge since the targeted beneficiaries are patients suffering from motor disorders and locked in due to paralyses. To make use of the motor imaginary signals for machine control, a BMI architecture with the combination of OPC (OLE process control), a standard communication protocol for the automation and machine control industry, is presented here. This model is designed and tested for a real-time communication between simulated actuators and MATLAB which was used for signal processing. The features of the motor imaginary signals are consolidated by Adaptive Recursive Bandpass filters and SFAM is used for classification. The implementation of the communication transfer with simulated actuators was comprehended with the help of the Matricon OPC as the Server and MATLAB OPC as Client. This system implementation gave a very promising result which is a confidence booster to assure the feasibility and effectiveness of implementing this method in physical setup for real-time applications with reference to the time of control response.

The motor imagery is one of the types of BMI in which the neuronal signals evoked from specific parts of the brain are sensed and translated into machine understandable controls in order to operate attached circuits of mechanical nature or electrical circuits with sensors or switches. BMI with motor imagery is regarded as a profitable system of support for people affected by disorders in their motor movements and disabilities like paralyses. Motor imagery is a well-identified method for enabling them to execute their desires of movement or any other form of activity presented to them under rehabilitation and assistance like activating their daily activities (e.g. operating home electronic equipment or wheelchair).

As specified in earlier chapters, the BMI comes under its two major divisions called invasive BMI and non-invasive BMI. An invasive BMI is usually carried out for clinical purposes that involve neurosurgical procedures for electrode implantation on the surface of brain which is known for its risks and always deemed dangerous and expensive too. On the other hand, the most popular use of non-invasive BMI involves wearable procedures for obtaining signals from the outer surface of the scalp through the electrodes embedded on the wearable. The EEG is a widely known and most preferable non-invasive technique for such signal acquisitions because of its non-invasiveness, cost-effectiveness, and easy implementation. The example presented in this section concentrates on the utilization of MRPs suitable for an EEG-based BMI. We can observe the MRPs usually in the EEG's mu-rhythm (8–12Hz) of a person during the performance or mere imagination of performing a motor activity. Here, we can see the use of the MRPs resulting from motor imagination which is also known as motor imaginary signals. These activities can be captured from the EEG's channels C3 and C4.

This BMI constructed using motor imagery potentials consists of four processes: the first step is the brain signal acquisition (the signal from C3 and C4 channels) signal preprocessing, (though the electrodes at C3 and C4 are used, they pick up some colored noise from other parts of brain) feature extraction (various techniques are used and some are discussed in this book), classification (wide range of classifications used and few are mentioned in this book), and translation into machine action. All these four steps have to be completed to obtain an efficient BMI.

As the output part of a BMI that controls a device or an actuator, the output of classification needs to be transformed into a control command and sent to a controller of the target device. At the other end, the receiving controller which could be a sensor that triggers the device or an uncomplicated programmable logic controller (PLC) operates the device to carry out the anticipated deed. In those earlier stages, these instructions were sent via an OPC (OLE process control) communication protocol, which was a standard data exchange methodology used in the automation and machine control activities.

Here is this example of a BMI using EEG; the design and demonstration is done to make it capable of sending control instructions via the OPC in a real-time communication system. In this demonstration an OPC simulator is used which simulates real-time connectivity to sensors or actuators. The raw EEG signals (not subjected to preprocessing) are acquired from the Graz BCI motor imagery datasets. As the next step, the obtained signals are subjected to preprocessing by means of bandpass filtering. So an adaptive recursive bandpass filter was deployed to extract the features which represent the motor imaginary movements from the EEG signals. The features of EEG which represent the motor imagery are then classified by a Simplified Fuzzy ARTMAP (SFAM) structure of Artificial Neural Network. These approaches were successfully experimented and tested in this work and implemented. As mentioned earlier, the classified signal outputs are interpreted and transferred to a simulated actuator connected via the OPC communication channel.

The deployed adaptive recursive bandpass filter always tracks the center frequency of the dominant signal of which the filter has only one filter coefficient that has to be updated, while the time function of the coefficient represents a distinct feature.

The SFAM used in this demonstration is for classification developed from a simplified version of Fuzzy ARTMAP (FAM). SFAM is higher in performance than FAM because of its potential to ease the computational overhead and architectural redundancy of the FAM network. Apart from that, it possesses the ability of exhibiting incremental supervised learning and high-speed training which are the essential characteristics of Machine Learning.

6.6.1 Demonstration Method

The work presented in this demonstration further shows the practicability and effectiveness of the novel signal processing methods in a real-time control system environment. The output communications based on the OPC standards were comprehended on a client/server environment in a wireless Local Area Network (LAN). As a performance measure, the response time calculated between the simulated actuator and the classification of the motor imaginary signals gives a clear advantage of this demonstration of BMI to be fit for a real-time environment.

6.6.2 Description of the Dataset

This method was experimented using the popular BCI Competition 2003, dataset III-b supplied by the university of Graz's Biomedical Engineering Institute, especially from Department of Medical Informatics. The EEG signals from the dataset were acquired through the EEG channels at positions C3, Cz, and C4. Later as a preprocessing, these signals were filtered by a bandpass filter for a frequency range of 0.5 to 30Hz, and it was sampled at 128 Hz. The person from whom the signals were acquired was in her 25th year of age and was relaxing on a chair with armrest during the signal recordings. This female subject was given a task to control a feedback bar by means of left-hand and right-hand movements totally by imagination. Guidance for what to imagine will be given to the subject for labeling the signals. The given experiment consists of 7 runs having 40 trials for each run. The experiment with all the seven runs was conducted on a single day. There were several minutes break in between each run was given to the subject. The dataset is separated into 2 halves and each half consists of 140 training and 140 test trials and comes to a total of 280 trials. The trials were conducted to have equal number of imaginations with right-hand and left-hand movements. Each trial was recorded for 9 seconds duration. At the third second of the recording, a visual clue, by displaying a pointing arrow, pointing to left or right to indicate which one of the motor movements is to be imagined, is given to the subject.

6.6.3 Preprocessing and Feature Extraction

The total number of 280 signals recorded from channels C3 and C4 were preprocessed by passing them through a bandpass filter in order. The filter used to do the bandpass is a seventh order Butterworth filter with a pass band of

9Hz with less than 1 dB of ripple. It was using a stop band of 11Hz with at least 6 dB of attenuation. The signals recorded from channel Cz were ignored since its contribution in discriminating the task was very less significant.

As a second step, the signals after preprocessing were fed into the adaptive recursive bandpass filter. The estimation of center frequencies of the dominant signals from channels C3 and C4 are done and tracked by the adaptive filter. The adaptive filter updates only one filter coefficient in order to adjust the center frequency of the filter bandpass to be matched with the input data. The filter used was a fourth order Butterworth bandpass filter as the adaptive filter where the filter function, T (z), is defined as

$$T(z) = \frac{D_0 + D_2 Z^{-2} + D_4 Z^{-4}}{1 + F_1 C(n) Z^{-1} + \left(F_2 C^2(n) + F_2'\right) Z^{-2} + F_3 C(n) Z^{-3} + F_4 Z^{-4}} \qquad (6.9)$$

where

$$D_2 = -2D_0,$$

$$D_1 = D_3 = -4D_0, \quad F_1 = -2l\left(2l + \sqrt{2}\right)D_0,$$

$$F_2 = 4l^2 D_0, \quad F_2' = 2\left(l^2 - 1\right)D_0,$$

$$F_3 = 2l\left(-2l + \sqrt{2}\right)D_0, \quad F_4 = \left(l^2 - \sqrt{2l} + 1\right)D_0,$$

$$l = \cot an\left(\pi BP\right)$$

The coefficient C (n) could be expressed as

$$C(n) = \frac{\cos(\pi\left(H_2(n) + H_1(n)\right))}{\cos\left(\pi BP\right)} \qquad (6.10)$$

where
$H_1(n)$ = normalized low cut off frequency,
$H_2(n)$ = normalized high cut off frequency,
BP = normalized bandwidth of the filter,
assumed constant.

Equations 6.9 and 6.10 give a clear picture that $C(n)$ is the only one coefficient updated by the adaptive filter due to the reason that it is also the only coefficient which is dependent on the center frequency, $(H_1(n)+H_2(n))/2$. Figure 3 shows the realization of the features for channel C3. The filter $T(z)$ self-adjusts itself to the center frequency of the input signal, by maximizing the output power of the filter. This is realized by introducing a standard gradient approach. The filter coefficient is then updated by using an algorithm called recursive maximum mean-squared. The time function of the updated coefficient is used as the feature that represents the signals resulted from right or left motor imaginary.

6.6.4 Classification of the Features

The SFAM used here in this demonstration is a supervised network architecture which has the ability to perform incremental supervised learning and high-speed training in fast learning mode. The SFAM architecture comprises of a Fuzzy ART (FA) module, category layer, and an Inter ART (IA) module. The category layers are connected via the IA module. The illustration of the architecture could be observed in Figure 6.18.

Classifications using SFAM consist of two stages, which are training the network and testing the network. As for the training phase, the input signal needs to be complemented so that it represents a pattern and is presented to the input layer of the FA module. Then the output classes of the category layer are presented with a binary value of 1 for the target class and 0 non-targets. Once this is done, mappings with the FA are created by the IA to classify

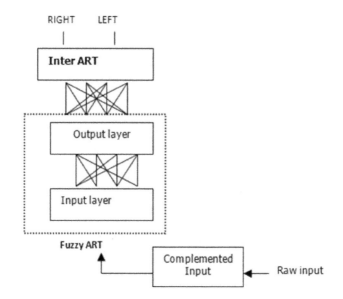

FIGURE 6.18
SFAM architecture

the outputs to different categories. In this case, a dynamic weight link will be established between many one-to-one mapping of the FA's output layer and category layer. One of the most important variables of the SFAM neural network is vigilance parameter (VP). The VP is a part of the IA module where its manipulation leads to the control of the granularity of the output node encoding. The high or low VP value determines fuzziness of the output node when deciding on how to encode the input patterns. The VP values range from 0 to 1 and the lower the VP the easier the matching criterion is for the output node.

In these experiments, the SFAM neural network was trained and tested using features from channels C3 and C4 separately. The 280 trials were first shuffled according to the seed values ranging from 1 to 10 where the seed value determines the order of the shuffle. {shuffle e.g. Seed1: 1,0,1,0,1,0,1 and Seed2: 0,1,0,1,0,1,0 and Seed3: 0,0,1,1,0,0,1,1, etc.}. This is done in order to result in better training performances of the network. After each shuffle, the trials were separated into two groups, where 140 trials were used for training and 140 trials were used for testing. Then each training feature vector is complemented in order to prevent category explosion and gives the networks its fuzzy nature. After this process, the features together with its labels are introduced to the FA for learning. The learning rate is set to fast, and with a VP of 0.20.

Once the network is trained, the test data are introduced to the FA to obtain categories from the output layer. The resulting categories are then used by the IA module to determine the responsible category layer node that refers to the predicted right or left-hand movement class. The classification outputs were translated, where right-hand movements indicate to switch on the motor, while left-hand movements indicate to switch off the motor.

6.6.5 Communication with OPC

OPC is an interface standard for remote real-time communication between a personal computer (PC) and control devices such as the PLC. The OPC uses a client/server mode based on Distributed Component Object Model (DCOM) for communications and data exchange. OPC also supports the network distributional application procedure communication as well as the application procedure communication on different platforms.

In this work, the Matricon OPC Simulator and MATLAB OPC toolbox was used. The Matricon OPC simulator simulates real-time communication environment between the actuators or sensors. This enables us to send instruction to a simulated controller or device without connection to a real device or controller. The MATLAB OPC toolbox was chosen because of its flexibility since MATLAB was also used for signal processing. The Matricon OPC interface forms a 'plug and play' environment between the MATLAB OPC client and Matricon OPC server. A standard interface was provided to the OPC objects by the Matricon OPC server component, which at the same time manages the objects. The MATLAB OPC client then creates and manages the server using the API provided by DCOM and accesses the server's data objects through the interface method. Matricon OPC server consists of three levels of objects, which are the server object, the group object, and the item object. The server object contains all information of the server and group object, while the group object contains all information of itself and the item object.

The MATLAB OPC client was used to send the classification output to the Matricon OPC Server via a wireless router within an Ethernet LAN. The OPC server then sends the intended instruction to the simulated actuator. The actuator writes the instruction to its memory, which represents an action of a mechanical or electrical device (e.g. motor, light switch).

6.6.6 Experimental Results and Discussion

Separate experiments were conducted with signal features from C3 and also C4. As mentioned in the previous section, the 280 trials were shuffled repeatedly for seed values ranging from 1 to 10. After each shuffle the dataset was divided into training and test set with 140 trials, respectively. This is done in order to obtain the best classification accuracies which depend on the order of features being presented for training and testing of SFAM. Table 6.6 shows the classification results for C3 and C4 respective to the shuffle seed values. From the results, we could observe that the classification percentage and time vary according to different seed values. We could also observe that the results for features from channel C4 give the highest percentage of 97% for seed value 2 while features from channel C3 give the highest percentage of 95% for seed value 2. Apart from that, the average classification percentage and time for C3 are 92% and 0.06s, respectively, while for C4 the average classification percentage and time are 94% and 0.05s, respectively.

The proposed BMI design was demonstrated using the signals from channel C4 with shuffle seed value 2 since this gives the best classification performances. This also indicates that signals from only one channel are sufficient for classification of motor imaginary tasks as mentioned by the researchers who have worked in this field earlier. The OPC simulated actuator chosen for this experiment was the bucket brigade with an integer type input value. Ten trials were chosen at random to observe the total response time between the classification and Matricon OPC's write process. The results are depicted in Tables 6.6, 6.7 and we could observe that the average classification time takes 0.0006 seconds while the write process takes 0.0055 seconds, with a total response time of 0.0061 seconds.

6.6.7 Consolidation of BMI Design

The work presented in this BMI example clearly demonstrates the BMI design using adaptive bandpass filtering and SFAM could be easily embedded with

TABLE 6.6

Classification percentage and time of signals from C3 and C4 respective to the shuffle seed values

Shuffle Seed	C3		C4	
	Classification (%)	Time (s)	Classification (%)	Time (s)
1	93	0.06	92	0.05
2	95	0.05	97	0.05
3	92	0.07	95	0.05
4	88	0.07	96	0.07
5	89	0.05	95	0.05
6	89	0.05	94	0.05
7	93	0.06	96	0.05
8	93	0.07	95	0.05
9	92	0.09	91	0.05
10	94	0.06	93	0.05
Average	**92**	**0.06**	**94**	**0.05**

TABLE 6.7

Response time between classification and OPC simulated actuator write time(s) for 10 trials

Trial	OPC write time(s)	Classification time(s)	Classification +PLC time(s)
1	0.0058	0.0028	0.0086
2	0.0030	0.0004	0.0034
3	0.0007	0.0004	0.0011
4	0.0152	0.0004	0.0156
5	0.0005	0.0004	0.0009
6	0.0005	0.0003	0.0008
7	0.0140	0.0004	0.0144
8	0.0005	0.0003	0.0008
9	0.0005	0.0003	0.0008
10	0.0147	0.0003	0.0151
Average time(s)	**0.0055**	**0.0006**	**0.0061**

the OPC communication protocol for machine control. The proposed method indicates that the real-time implementation can be achieved between signal processing using MATLAB and any mechanical or electrical actuator through the OPC server and the test results prove the effectiveness and feasibility of the method.

Furthermore, this method enables us to obtain classification results by using features from only one channel at a time, which is suitable to be modeled for real-time application. This technique can be also extended in identifying other useful sources and components buried in the recorded EEG signal. Apart from helping the disabled, the presented method can be also used to realize advanced control in an industrial environment, since the usage of OPC communication standards are popular in those environments (Figures 6.18, 6.19, 6.20, 6.21).

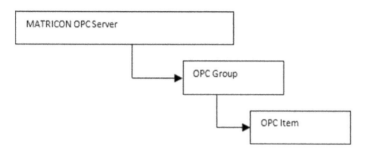

FIGURE 6.19
OPC System hierarchy

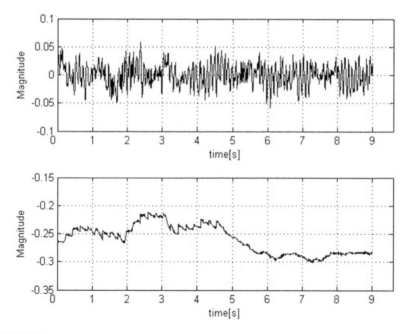

FIGURE 6.20
Average of 10 trial realizations of signals from channel C4 (Top). Mean values of the extracted features (Down)

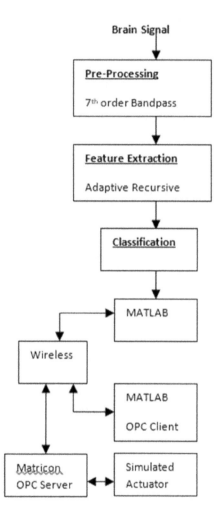

FIGURE 6.21
Proposed BMI architecture

6.6.8 BCI Example 6b: Motor Imagery – Fractal Dimensions in Estimating Features

The same motor imagery-based BCI by means of brain activity into computer commands, thus providing nonmuscular interaction with the environment, can use various feature extraction tools such as fractal dimensions. The sensorimotor rhythms (SMRs) are rhythmic brain waves found in the frequency range of 8 to 12 Hz over the left and right sensorimotor cortices of the human brain. By the results of various researches in the area of motor imagery reflections on the brain, one can understand that any form of movement, preparation of movement, and motor imagery desynchronize SMRs, whereas during relaxation or post movement, they are synchronized. The reason for using motor imagery-regulated SMRs in BCI is that they do not require any muscular activity. As we have seen in previous sections, this is beneficial for people with neurological disorders, since they can't produce any muscular activities. One more advantage is the short training period it requires for use in BCI. The basic idea of such BCI using motor imagery tasks is to identify the synchronization and desynchronization of SMRs using the brain signals. Researchers used the imagery of hand movements, foot movements, and tongue movements in many of the systems developed for this purpose. The SMRs need to undergo preprocessing, feature extraction, and classification operations after their successful acquiring from the brain during the imagery activity.

The difference between VEP-based BCI and Motor Imagery-based BCI are in the time of signal acquisition. Features are found in VEP-based BCI after the onset of the visual stimuli whereas in motor imagery it is found during the imagery. Feature extraction or feature detection is done by means of simplifying accurately the exact representation of event data by removing all artifacts, noise by reducing the signal dimensionality. The more we eliminate the noise without eliminating the feature components, the more easy is the classification. If the classification is carried out without an accurate feature extraction process on the raw signals which are usually of a high dimension and redundant data, its computational complexity would be high and result in inaccurate results. The tool used here in this experiment is Fractal dimension which is a statistical function that visualizes the complexity of a presented object or a quantity that is self-similar over some region of space or time interval. It is a well-known function successfully tested and found suitable in various applications of wide varieties of domains in characterization. So it is suggested to deploy it in motor imagery-based BCI and found fruitful by the obtained results. But the fractal dimension estimation is not a singular method. It has several functions to estimate fractal dimensions, and some of which are more suitable than the other to some specific applications. It is necessary to choose a most suitable method of FD estimation in order to achieve a higher classification accuracy and speed.

In this example, four methods namely Katz's method, Higuchi's method, the rescaled range method, and Renyi's entropy were used for feature extraction in motor imagery-based BCI by conducting an offline analyses of a two-class motor imagery dataset. Fuzzy k nearest neighbors (FKNN), support vector machine (SVM), and linear discriminant analysis (LDA) classification were tested in combination with these methods to determine the best methodology on performance. As the next phase, the methodology was modified by implementing time-dependent fractal dimension (TDFD), differential fractal dimension (DFD), and differential signals (DS).

6.6.8.1 Data and Method of Operation

The BCI Competition II (Dataset III) was taken for operation since the benchmark datasets are the proofs to the performance of suggested methodology and easy to compare with the results of other works. This dataset was provided by the Department of Medical Informatics, Institute for Biomedical Engineering, University of Technology Graz, and commonly known as Graz dataset. The data of this dataset was acquired over seven runs from a healthy 25-year-old female subject during the imagery of right- and left-hand movements. The captured signals during the imagery were recorded with a sampling rate of 128 Hz from all three electrodes placed at the standard positions of the 10–20 international system (C3, Cz, and C4) and filtered for removing noise between 0.5 and 30 Hz. Each run consisted of 40 trials and each trial lasts for nine seconds duration. During the first two seconds of each trial, neither a stimulus was presented nor did the subject perform any motor imagery task. After this period, an acoustic and a visual stimulus indicating the beginning of the motor imagery task were presented to the subject. Consequently, during the next six seconds, a cue (a left or right arrow) indicating the required motor imagery task was presented (in a random order for each trial), and the subject followed the direction and performed the task. A feedback bar was displayed during this time. There were a total of 140 samples each in the training and testing sets. As mentioned, the preprocessing was done by a zero phase filtering using a sixth-order bandpass digital Butterworth filter with cutoff frequencies of 0.5 and 30 Hz in both in the forward and reverse directions. Each trial was extracted for the last six seconds to discard the period without any motor imagery. Two different electrode configurations (C3 and C4, and C3, Cz, and C4) were tested in this experiment.

6.6.8.2 Feature Extraction by Assorted Methods of FD

In Katz's method, Higuchi's method, and the R/S method, the fractal dimension of the samples from selected electrodes was concatenated into feature vectors. In the TDFD, DFD, and DS methods, the fractal dimensions were estimated using the fractal dimension estimation method of the methodology with the best performance.

In Katz's method, calculation of fractal dimension of a sample is by the sum and average of the Euclidean distances between the successive points of the sample (L and a, resp.) and the maximum distance between the first point and any other point of the sample 'd' was also calculated.

$$D = \log(L/a) = \log(n)$$

$$\log(d/a)\log n + \log(d/L)$$

where L is n divided by a.

In Higuchi's method, the calculation of FD of a sample is done as follows: first, subsample sets (X_k) are constructed from the samples (X) as

$$X_k^m = \left\{ x\left(m + ik \right) \right\}_{i=0}^{((N-m)/k)},$$

Finally, the fractal dimension of the sample (D) is solved from

$$\langle L(K) \rangle \alpha k^{-D}$$

where L is the average of L_m. Three K_{max} values from the range of 8–18 and 8, 13, and 18 were tested.

In R/S method the FD is calculated for a sample by iteratively dividing it into non-overlapping subsamples with decreasing subsample size and performing the following operations at each iteration: for each subsample, a new subsample (X) is constructed from its zero mean (ξ) such that the nth point of X is the cumulative sum of the first n points of ξ. Then, the difference between the maximum and the minimum values and the standard deviation of X (R and S, resp.) are calculated in order to obtain their ratio (R/S). Finally, R/X of each X is averaged (R/X_{avg}). After obtaining R/X_{avg} at each iteration, the Hurst exponent (H) becomes the slope of the log–log plot of R/X_{avg} versus subsample size. The fractal dimension then becomes 2-H.

In Renyi's Entropy, FD is calculated as generalization of Shannon's entropy. Renyi's entropy is defined as

$$Rq = \frac{1}{1-q} \log 2 \left(\sum_{i=1}^{n} p_i^q \right)$$

where q>0

6.6.8.3 TDFD Method

In TDFD method, a window (with size s) is slid over a sample by a time step, and the fractal dimension of the part of the sample inside the window is estimated. The fractal dimensions were concatenated into feature vectors. Different window sizes were tested using a time step of one second.

6.6.8.4 DFD and DS Methods

The DFD method is a variation of the DS method. In the DFD method, first, the fractal dimensions of the samples from selected electrodes are estimated, and then, the pairwise differences of the fractal dimensions are calculated.

However, in the DS method, first, the pairwise differences of the samples from selected electrodes are calculated, and then, the fractal dimensions of the pairwise differences are estimated. In both methods, the resultant values were concatenated into feature vectors. Only the three electrode configuration was tested, since the two electrode configuration results in one-dimensional feature vectors.

6.6.8.5 Classification

After a careful construction of the feature vectors, using specific FD methods the test samples were subjected for classification as imagery left- or right-hand movements using different classifiers namely FKNN, SVM, and LDA.

Fuzzy KNN is a variation of KNN. The main difference between the two is that KNN assigns a class label to a sample that is most frequent among the k nearest neighbors of that sample, whereas F-KNN assigns a membership value for each class in this neighborhood and classifies the sample as the class with the highest membership value. The membership value for a class was calculated by dividing the sum of the distances between the samples belonging to this class and the test sample by the sum of the distances between all the samples in the neighborhood and the testing sample. Number of nearest neighbors between one and the square root of the sample length were tested.

SVM separates the samples using a hyperplane that maximizes the margin between those belonging to different classes. SVM with a linear kernel was used.

LDA finds a linear combination of features that best separates the samples belonging to different classes and can be used as a classifier. To assign a class label to a sample, the probabilities of the sample belonging to each class were estimated using LDA. The label of the class with the highest probability was then assigned to the sample.

6.6.8.6 Findings

The classification accuracies and the computation times were evaluated for each fractal dimension calculation method and classifier combination. Katz's method was the fastest method, and combining it with FKNN, the highest classification accuracy of 85% (the three electrode configuration and k=9) and the second highest classification accuracy of 83% (the two electrode configuration and k=9) were achieved. Method with any classifier performed the slowest with the classification accuracies and the computation times ranging from 69% to 71% and 7.32 to 11.07 s, respectively. On the other hand, Renyi's entropy with any classifier performed the worst with

TABLE 6.8

Comparison of the computation times and classification accuracies

	TDFD method C3, C4 C3, Cz, C4	DFD method C3, Cz, C4	DS method C3, Cz, C4
Classification accuracy (%)88	(k = 5, s = 10) 85 (k = 7, s = 64)	84 (k = 11)	1 (k = 11)
Computation time (s)3.47	(k = 5, s = 10) 0.94 (k = 7, s = 64)	0.41 (k = 11)	0.26 (k= 11)

the classification accuracies and the computation times ranging from 55% to 66% and 1.84 to 4.87 s, respectively. The performances of the rest of the combinations were similar. The classification accuracies (except for the R/S method and Renyi's entropy) and computation times (except for the R/S method) increased with the number of the number of selected electrodes.

Table 6.8 shows the computation times and classification accuracies obtained by modifying the best-performing methodology. Although all the modifications increased the computation time, further improvements in the classification accuracy (by 3%) were achieved only by implementing TDFD method (the two-channel configuration, k = 5 and s = 10). However, implementing the DFD and DS methods resulted in lower classification accuracies.

Mental activity may modulate FD of EEG signal which implies that it is timed-dependent in nature. By implementing TDFD method in Katz's Method with FKNN, we may measure the fractality in short time intervals of time-sequential data from one end of the waveform to the other sequentially, and we may observe the dynamical changes in the FDs with respect to the time series. These FDs, namely, are referred to the time-dependent fractal dimensions (TDFD).

Katz's algorithm is the most consistent method due to its exponential transformation of FD values and relative insensitivity to noise. Hiaguchi's method, however, yields a more accurate estimation of signal FD when tested on synthetic data, but it is more sensitive to noise. In the experiment, EEG datasets used are real datasets which contain noise; hence, Katz's method exhibits better result.

In conclusions, as stated before in the beginning of this experiment sample, all fractal dimension estimation methods are not appropriate to all types of data exhibiting fractal properties, commonly used fractal dimension estimation methods to characterize time series with different classifiers were evaluated to find the most suitable method for motor imagery data. Katz's method with FKNN was determined to be the best methodology, and the results were further improved by implementing the TDFD method. The results give proof to further research to use this methodology in online analysis of motor imagery data and analysis of other signals.

6.7 Sign Language

6.7.1 Need for Sign Language

Sign language is the main mode of communication for deaf people, they still experience problem in communicating with common people as everyone does not understand sign languages (Figure 6.22).

Originated from the areas of the United States of America and Canada, the mostly accepted and widely used around the globe is the American Sign Language (ASL). No one invented it with any formal grammars or rules. It evolved as a practice among groups of people who have problems with their conventional communication. The very reason for its wide usage is due to its completeness, and it naturally resembles the same linguistic properties as spoken languages. But the grammar of ASL is slightly different from English. ASL is a combination of expressions by movements of hands and face. Another uniqueness of this ASL is that the people who are deaf and hard of hearing, as well as people with normal hearing and speech are also using it.

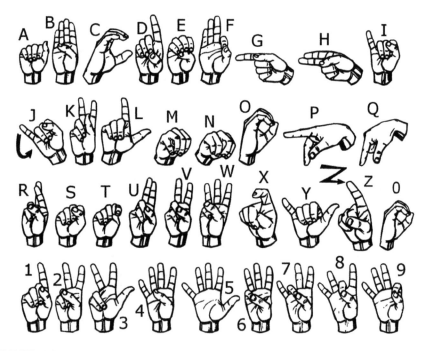

FIGURE 6.22
American sign language alphabets and numbers

Children are taught by their parents in the beginning, and after this the children pick up more vocabulary and accents of ASL from schools and their community similar to any other spoken languages.

Researchers like Cemil Oz and Ming C Leu say that the communication problem of deaf people still exists even after the sign language practices, especially when communicating with people who hear normally, since, almost all of them do not understand sign language system. So it is felt by research teams from another area to use biosignals extracted from hands while the sign language is performed, as an effective mode of communication that is parked under the rehabilitation engineering field. As a notable example, the sign language, which is a highly visual–spatial, linguistically complete, and natural language, is the main mode of communication among deaf people and is automated for both ways of understanding. The enhancement to the system helps to include normal people too in finding an easy way to communicate with the mute.

A complete formal set of communication through sign language demands strain of fingers, training, accuracy in differentiation, and complex finger twists. That is why it is emphasized to expose children to the outside world as earlier as a child begins to acquire language. This will lead to a better language, cognitive, and social development skills in children. Many researchers found that the first few years of child's life are the most important for the development of language skills, and even the early months of life can be crucial for establishing successful communication with caregivers. But a single sign language is not used globally, and it varies according to the nations. Adopting ASL for communicating with patients is carried out by caretakers since it covers all categories of patients and acts as a common medium in communication. But it suffers from overlaps, extensive training period and again leads to fatigue. An alternative solution which simplifies the drawbacks of ASL is the gesture-based communications. Use of the commonly understandable gestures with corresponding meaning is highly essential at this point of time (Figure 6.23).

6.7.2 Sign Language Automation

As a detailed exploration into this research area, a robust two-way communication system is found necessary for helping in identifying the best possible method of communication. The objective is to provide a solution for hearing not only to the impaired or mute but also to the commons who wish to communicate with their challenged counterparts. The gesture communication is the automatic communicative system for common environment. It is proved by two experiments of this area done by Christian Obermeier and team in 2012 that the speech and gesture can be integrated. In one experiment, the normal hearing community is allowed to view videos in which gestures are made and the voice is buffered due to noise. The target word is found and

FIGURE 6.23
Experiments of automating the sign language using wearable hand gloves

the delay of buffering is recognized. As a second experiment, the same video is viewed by the hearing impaired and age matched people. At the end of the experiments, it is found that the speech gesture ambiguity is easily identified only by the hearing impaired. Both the experiments indicate that gestures are more beneficial for communication despite the reason for the disabilities are either external or internal. Sign languages are used by the hearing or speech impaired for their communication. They use their hands and gestures for the communication more easily than words.

6.7.3 Classes of Sign Language Automation

There are parallel research and development for sign language recognition systems is going on under three major categories. They are broadly classified as

(i)   Computer-vision-based recognition
(ii) Data-glove and motion-sensor-based recognition, and
(iii) A combination of these two methods.

Allen M.J. and team categorized these three methods in 2003, and a computer-vision-based ASL recognition relies on image processing and feature extraction techniques for capturing and classifying body movements and hand shapes when a deaf person makes an ASL sign. On the other hand, data-glove and motion-tracker-based ASL recognition methods use a sensory glove and a motion tracker for detecting hand shapes and body movements. The third method includes a combination of techniques from these two methods.

New research developments continued to remove barriers for people who are speech and hearing impaired by using artificial intelligence. Both hardware and software are developed that improve the knowledge of deaf people in communicating. Sign languages are used for better communication for the blind. Research also continues to analyze speech and convert it either to sign or text for deaf people.

Modern research by Grobel. K. and Assan. M., Mahmoud M. Zaki and Samir I. Shaheen has seen the development of computers that can recognize sign language. These are carried out with components as the hand shape, orientation, and movements, which incorporate virtual reality sensors to capture both isolated and continuous hand signs. The skin color, dominant hand tracker, component locator and the Markov Modeling, and neural network systems have made a significant impact in this regard.

6.7.4 The Hand Glove or Data Glove

Data glove (commercial name: 5DT Data Glove 5 Ultra) is a device similar to the conventional glove and it is worn on the hand. The data acquired from the data glove is a 14 column matrix since 5DT Data Glove has 14 electrode channels in it.

Ultra model has 14 fully enclosed fiber optic bend sensors embedded in it.

These bend sensors were spread in such a way that one sensor is for knuckle and one for first joint abduction, and sensors between fingers of the hand. It is designed to satisfy the requirements of modern motion capture and animation professionals. An assorted software support is given in data glove. The bend sensors are capable of producing continuous data samples with flexure resolution of 12 bit A/D / with an USB interface using kaydara MOCAP, Discreet 3D studio Max, Alias Maya, SofImage XSI, SDK, and Glove.

Manager Utility Software. The sampling rate of this glove is minimum 75Hz. This glove is available in most of the laboratories working with rehabilitation, assistive technology, robotic communication, cyborg, etc.

The data acquisition done through this glove is represented in 8-bit flexure resolution, at the sampling rate of minimum 75Hz (5DT Data Glove 5 Ultra). Originally the data glove is designed as a 3D input device, and this makes it suitable for a wide range of applications like control and manipulation of virtual worlds, gesture and cognitive media, physiotherapy, rehabilitation, and control device for artists in remote controlled environments.

The glove can be used in either one of our hands for capturing hand dynamics. The data glove is used for acquiring signals produced during the various gestures of hand. The reason the data is called as signal is due to its continuity and variance. The data glove is efficient and facilitates sensing and fine-motion control in robotics with the varying degree of the sensors placed in it. Reinhard Genter and Joseph Classen 2009 say that the use of data glove is a new dimension in the field of medicine and health care for rehabilitating. The widely available and common input devices are restricted only to certain purposes, but the data glove offers multiple degrees of freedom for each finger and hand as well. This permits the user to communicate to the computer or machine to greater extent than most other input devices.

During the data acquisition in research projects, the interfacing of data glove with the computer is done through a cable to the platform independent USB port or through wireless by means of Bluetooth technology with up to 20 meters of distance. This glove is fabricated with the flexible black stretch lycra material to fit to many different hand sizes unreservedly. The data gloves are used for many purposes like communication, forgery detection, signature verification, etc., and here it is used for monitoring and help in the communication of the elderly and disabled with the caregivers or health aides.

6.7.5 Advantages in Glove-Based Sign Language System

The basic architecture of a glove-based (signal based) system consists of a data glove with embedded flex sensors, worn on a hand while making signs and movements for the sign language communication. The recent glove, called a data glove or the cyber glove, incorporates between 18 and 22 sensors and is connected to a computer through a serial cable. The signs that are produced by the finger movements are captured as signal group by the computer. The sensor positions and movements of the glove are interpreted as the data of the hand movements representing the sign language alphabet and are wirelessly transmitted through a portable device to the computer and from where it is displayed on the computer. A supporting software system records these signals and converts them to be expressed in written or spoken form of the language for communication.

A noteworthy problem in such a developing field is the lack of uniformity in the sign language used worldwide. Sign languages are not international, each country uses its local sign language. There is no common mechanism for translation of these languages. American Sign Language (ASL), for example, is not simply a visual representation of English, but has its own specific rules and grammar. The signed system and signed English are easier; however, developing a software for transliteration of a combination of signed language actions into a natural language like English is not simple and needs a systematic processing. Signing is a duel process and requires both the ability or skill to read signs and the skill to make or render signs. Some systems developed by Cemil Oz and Ming C. Leu for the recognition of ASL words are done with the Artificial Neural Networks (ANN) and translate it into English. The researchers use a sensory glove, cyber glove, and flock of birds' 3D motion tracker to extract the gesture features. This sign capturing device is set up with two neural networks. The velocity network – recognizes the duration of words with the hand speed; word recognition network – classify the ASL signs into words. Researchers like C. Shahabi used the cyber glove to recognize the hand signs and gestures from the acquired haptic data.

Human–computer interaction technology also has a numerous applications of wearable gloves where sign language, movement control, and gesture recognition are the important interaction methods and are found appropriate among all available technologies. The recolonization systems of sign language, the data glove with bend sensors, gyroscopes, and the accelerometers were chosen by the researchers and the data glove with the accelerometers are used to segment the different signs performed by the users and then the segmented values are classified for the real-time environment.

As recent development, wireless Bluetooth data gloves for sign language recognition system are used for assisting the speech impaired population in communicating with the common people. This human–computer interaction technique converts signs into speech (sound) so that the deaf people can communicate with the common people. A system developed by Tan Tian swee in 2007 using the Hidden Markov Models for this purpose recognizes 25 common words from the Bahasa Ishyarat Malaysia (BIM).

The data from the hand while performing signs is passed to the connected computer system as continuous values through the Bluetooth. The sign language analysis is done with the arm joints with a camera computer vision algorithm. This technique was carried out by the researchers to avoid the vision-based vagueness.

These sign languages not only map the letters, words, and expressions of certain languages to a set of gestures performed but also recognize the two-handed signs generated and them. This was proved by an Arabic sign language dictionary recolonization system developed by Mohamed A Mohandes in 2010 – 2013 with the help of a set of cyber gloves.

Cemil Oz and Ming C. Leu in 2011, using a cyber glove (data glove) integrated a system and tracked the movements of hands and are traced by a Flock of Birds 3-D motion tracker to extract the gesture features. The usage of these motion trackers is expensive and ASL should be taught prior to the elderly and the disabled that is difficult.

All these systems are undoubtfully a breakthrough in the rehabilitation of mute people and significantly help them in using their skills of communication to a better extent and to a wider audience. But there are some limitations to this sign language automation system to be used as an emergency response system due to its requirements for precision, uniformity, and stability.

A requirement for a simpler communication interface for people with major disability, sickness which causes disability to communicate normally, and deaf people while they are sick is felt a lot in clinical practices. Even for normal people it is difficult to communicate during emergency, with mask, with oxygen mask, during unbearable pain, and while feeling difficulty in breathing. An efficient interface that can understand simple gestures of patients during these difficulties is found necessary for the medical personnel and the caretakers. The monitoring process of elderly during clinical process involves compulsory and temporal but continuous communication among the patient and their caretakers. Miscommunication will lead to adversative things. A simple wearable system other than sign language grammar could interpret the implicit message to the caretakers or to an automated lifesaving device.

Another interesting work is combining the photometric signals of a person along with the sign language gestures shown to reinforce and make it more distinct and unique for the use of more than the purpose of communication.

This system uses an American Sign Language which is the most common sign language used by deaf communities all over the world. The sign language becomes a must nowadays among the communities whereever the deaf people exist. The complex spatial grammars of sign language are distinctly weird from the grammars of conventional spoken languages. A few sign languages have obtained some form of legal recognition, while others have no status at all. In some countries ASL is mixed with other indigenous sign languages. The countries where American Sign Language is used or commonly included are Malaysia, Kenya, Philippines, Nigeria, Chad, Haiti, Central African Republic, India, Singapore, Zimbabwe, and Burkina Faso (Figure 6.24).

Enhancement to the sign language system has been suggested by many researchers like Plamondan, Srihari, S. Rahee, B.H. Yang, H.H. Asada, Y.Y. Gu,

FIGURE 6.24
Teachers using sign language to teach hearing impaired students in India

Y. Zhang, and Y.T. Zhang and biosignal-based security features are also considered as a unique alternative for applications that require some password evidence like banking or e-commerce. Making use of multimodal biometric technology that gives unique and robust identification features for every individual is in great demand nowadays for security environments that require high-quality authentication methods. Using PPG waveforms to distinguish individuals using their heart rate biometric component was suggested by researchers in 2007 [J.C. Yao et al.] and is employed in protected applications, e-transactions, and access control mechanism. In addition, combination of using data glove signals and PPG at the same time based on signature verification system has been demanded in 2010, which you will see in Chapter 7.

6.8 Wearables in Treatment

Every patient, even sometimes the caregiver, needs to wear different kinds of clinical wearables when undergoing treatment in hospitals nowadays. Starting from pulse oximeter to life support systems, ventilators for respiration system, etc....comes under the wearables essential for treatments.

Worldwide one person every two seconds suffers a stroke. Stroke is the third leading cause of death in the United States. More than 140,000 people die each year from stroke in the US. Each year, approximately 795,000 people suffer a stroke. In Canada, in 2000, stroke accounted for 7% of all deaths – 15,409. In India, Southern India found that 25% of stroke patients were less than 40 years of age. Cerebral Venous Thrombosis was found to be 12 times more common in India than in Western countries.

In the whole World, according to the WHO, 15 million people suffer stroke worldwide each year. 5 million die and another 5 million are permanently disabled. Europe averages approximately 650,000 stroke deaths per year. According to 2011 census 11,79,963 disabled are in Tamil Nadu.

According to Census in 2001, 21 million people in India are having disabilities in which movement disabilities are 6,105,4770.6, with 2,263,821 mental disabilities and 1,640,868 speech disabilities (Source: Home ministry of India). The total number of disabled people in India is around 1 billion people, which is 15% of the world's population (Source: WHO and the World Bank).

The devices that are manufactured are not only limited to the patients but also help physiotherapists, occupational therapists, and speech therapists who work with stroke-affected and disabled people. Efforts are taken to accommodate all these care and treatment givers in the application of wearable devices so that their stress and working difficulties get reduced through these devices. Another reason for the use of such devices is based on the statistics that at least 16% of the population worldwide will suffer a stroke in their lifetime. It is calculated that every four seconds in the world someone dies as a result of stroke. Up to 90% of strokes are preventable if the risk factors are managed appropriately.

Wearable devices for all types of poststroke patients, postoperative patients, patients having difficulty in breathing, premature infants, psychiatric patients, patients undergoing operation, dialysis, fracture recovery, gastroenteritis corrections, and other chronic diseases are essential.

These devices are also useful for children who are not able to access regular and proper services like special education, physiotherapy, occupational therapy, or communication therapy on a daily basis and for those who need extra care in special schools/centers.

6.9 The Wearable Assistive Device PhysiofastHeal

A device invented to automate the process of physiotherapy is named as PhysiofastHeal.

6.9.1 The Need and Challenge for the Invention of the Device

This device 'physiofastHeal' was developed as an outcome of a personalized encounter with a patient recovering from a spinal card injury. The increasing demand for human productivity especially the independence and freedom of action for poststroke persons, differently-abled persons and special children are in focus nowadays. Having it in mind, this novel assistive technology is proposed to succor the patients, caregivers and the therapists (Figure 6.25).

6.9.2 Beneficiaries

Post Bone/Muscle injury Patients can recover fast by giving a timely and proper physiotherapy exercises and rejuvenating treatments by physio-fastHeal device. Patients breaking their limbs have to undergo an intense series of physiotherapy. Apart from accident injury, physiotherapy is an effective treatment for aged, special children with cerebral palsy, etc. This device is aimed to be suitable for persons of any age affected by Cerebral Palsy, poststroke patients, postoperative/fracture patients, and also covering the requirements of people involving in gymnastics, body builders/instructors, dancers/dance masters, drill students, police, army, and other forces. It is entirely different and cost-effective from all the costly wearable gadgets available in market at present. It is designed to be a cheaper, reusable, interchangeable, wearable attire suitable for the limbs (left leg, right leg, left hand, right hand, neck).

6.9.3 System Design

This physiofastHeal was developed with a set of handmade varying pressure and bend gauges embedded in the elastic tube-like stretchable cloth. PhysiofastHeal is an open-ended product and can be adjusted and modified according to the requirements that vary from person to person. The idea behind this invention is to monitor the given Physiotherapy treatment as accurate as the physiotherapist initiated/taught. It is to monitor and cross-check the frequency, the level of correctness, and the duration. All these parameters will be recorded. The instant and history feedbacks (in visual and sound) will be given through the attached hardware using Bluetooth technology. Apart from that, guided therapy programs can help the user to carry out the therapy without boredom using simple tasks, color codes that enable the users to interrelate color and the pressure associated. This device helps them to follow color codes or interesting paradigms to complete their physiotherapy easily and perfectly by giving them varieties of feedback.

In addition to that, it gives a session report to the therapists to ascertain the quality of each session and presents a readily available report for the therapist to decide improvements, alterations or next level of treatments.

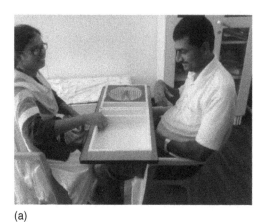

(a)

(b)

FIGURE 6.25
Nailing practice for recovery from spinal card injury

Thus, physiofastHeal is made to align the physiotherapy activity as per the clinical standards. This system is designed as an expert way to insist for the right therapy to recover the users as per the best standards in clinical practice.

The PhysioPowerHeal, next version of PysiofastHeal also calculates the performance via the Bluetooth files periodically, thus preserving the performance of the subject by giving feedback in both directions using the associated S/W App created.

6.9.4 Technology Design

This technology makes it very unique by the fusion of Assistive technology with wearable technology. Parents and caretakers of such stroke affected people, Children suffering from CP, post-injury patients, and aged are struggling to help and unable to nourish them every day and are always looking for help from technical people. According to a mini survey taken by me in physiotherapy centers, hospitals, normal, and special schools around Salem and Namakkal districts of Tamil Nadu, there is a quite significant amount of stoke patients of different age group and special children especially affected with Cerebral Palsy (CP) are looking for an assistive technology to maintain and or reequip their health particularly on motor activities in limbs and neck. This novel idea is to infuse Wearable Technology into assistive technology by Precision Aided-Physiotherapy which is essential to those with all the above categories to:

- Improve range of movement • improve strength • improve balance • improve mobility •improve independence at the earliest recovery time.

Performance and Advantages
The physiofastHeal is designed and fabricated to,

1. Eliminate the human dependent activities in treatments by automation, and
2. Reduce the boredom of repetitive conventional therapy in an entertaining way.

This product is very much essential to

1) Secondary stroke prevention
2) Improvement in Activities of daily living
3) Improved Mobility
4) Spasticity
5) Reduction in Pain

6) Incontinence

7) Communication

8) Keeping up good Mood

9) Active Cognition

10) Happy Life after stroke

11) Proper Relationship with family

The benefits of wearable device physiofastHeal include the following: cost-effective but efficient care, faster decreasing of pain by usage, steady improvement in joint mobility, quick increasing strength and coordination, improved cardio-respiratory function, less transportation efforts for treatment, easy diagnosis by medical personnel, and enhanced record of history of treatment.

6.9.5 The Innovation Introduced in PhysiofastHeal

The presented wearable product (mentioned in Section 6.9.4) is entirely different from all the available costly wearable gadgets. It is designed to be a cheaper, reusable, interchangeable, wearable attire suitable for the limbs (left leg, right leg, left hand, right hand, neck). This product will be developed with a set of handmade varying pressure and bend gauges embedded in the elastic tube-like stretchable cloth. This product is open-ended and can be adjusted and modified according to the requirements that vary from person to person. The idea behind this product is to monitor the given physiotherapy treatment as accurate as the physiotherapist initiated/taught. The frequency, the level of correctness, and the duration all will be recorded and immediate and history feedbacks (in visual, sound) will be given through the attached hardware via Bluetooth technology. Apart from that, guided therapy programs can help the user to carry out the therapy without boredom using simple tasks, color codes that enable the users to interrelate color, and the pressure associated. This device helps them to follow color codes or interesting paradigms to complete their physiotherapy easily and perfectly by giving them varieties of feedback. Finally, it gives a session report to the therapists to ascertain the quality of each session and gives a readily available report for them to decide improvements, alterations, or next level of treatments. Thus, this device is made to align the physiotherapy activity as per the clinical standards. The system is designed as an expert way to insist for the right therapy to recover the users as per the best standards in clinical practice. The proposed product also calculates the performance via the Bluetooth files periodically, thus preserving the performance of the subject by giving feedback in both directions using the associated S/W App created.

6.9.6 The Uniqueness of Innovation in Comparison to the Others in This Sector

The most relevant available technology to the proposed work is two in categories. One issuing the movements converted into predefined words or statement by Professor Prabhatranjan (http://www.ranjan.in/?cat=75) on a mobile application enabled a CP child to communicate her basic needs through just one touch. In the second stage, the 9-year-old was introduced to the neuro-headset. It had eight basic icons. This could be upgraded to more icons based on a user's needs. The icons had symbols for basic needs like food, water, washroom, lights, and one option to call the mother. The icons could be linked to voice message recorded in any language. The icons could also be set on auto-scroll for people who have very little control over their limb movements. This invention is purely on a customized basis. The next is by Professor Anil Prabhakar of IITM, (http://www.ee.iitm.ac.in/2015/03/intelligent-gesture-to-speech-device/) with the help of a gyroscope and an accelerometer pair embedded on a wearable casket like a watch. This device involves multi-dimensional data and helps the special people with the android app to communicate with predefined words or sentences. The early intervention by clinical procedures is important for any ailment to the patients. This helps fast recovery and avoids further complications and deterioration. The advantage of this PhysiofastHeal is tested over multiple experiments in other previous works (data glove used for signature verification, sign language, emergency response, etc.). The data processing is much simpler by less complexity. The PhysiofastHeal is much simpler and cost-effective than a data glove (imported). The PhysiofastHeal covers a wide range of users and applications (poststroke, post-injury. Cerebral Palsy, Aged, and Body builders). The PhysiofastHeal is an open architectural one which can make the physiotherapy activity fun and easy by guiding the user by different color codes, numbers, or directions (paradigms). The same device can be overloaded with multiple activities and feedback (by varying only the S/W without altering the hardware setup) which is a most remarkable advantage of the product. This avoids tiredness and boredom caused by monotonous action to the users. On the other side more advantage comes to the caretakers or the treatment givers (ortho, Physiotherapist) in the form of report feedback. The feedback and analysis of the performance of user in every session or single activity is the best clinical part of the system, which helps the therapists and doctors to decide on treatment intensity and modifications. Finally, it is much cheaper and depends on no import requirements, thus making the product much cheaper than any other devices available in the sector.

The validation of this proposed device and its technique has been discussed in parts with various experts of the domain and tested with preliminary lab experiments. This new device PhysiofastHeal is designed with much simple technology, thus reducing the processing requirements and cost involved.

Moreover, PhysiofastHeal is an open-ended device to make it possible to be used for multiple activities on the therapeutic paradigms and suitable for many ranges of people who require treatments. The PhysiofastHeal technology uses simple flex and pressure sensors conveniently made for the very specific purpose with much ergonomic care and helps the users to use it naturally like any other dress or wearable. (Figure 6.26) (Table 6.9).

The benefits of physiofast-heal are as follows:

- Occupies less space
- Bluetooth controller
- Lightweight
- Portable
- Low cost
- Durable

FIGURE 6.26
Analyzing the normal physiotherapy given by the therapist

TABLE 6.9

Impact on stake holders

Stakeholder category	Advantages & Response of the Stakeholder
Stakeholder 1: Patients	• Easy, entertaining, educating, and innovative way of performing exercises in a prescribed and clinically accepted way • Results in completion of exercise schedules as defined and timely • Results in fast and precise recovery 'why you saying it now and where were you when I was in hospital' Gopi (22)
Stakeholder 2: Caregivers	• Freedom from being an equal involver while the patient is doing physio. • Need not to give instructions, counting, and can move to other work until the patient finished a routine.
Stakeholder 3: Clinical Experts	• Freedom from listening to every case by case history of treatment. • Helps in very fast decision making in altering the treatment and levels. • Covering more patients in the available time. 'Give the device to me immediately' Mrs. P. Usharani, physiotherapist (38) Ecomwell orthopedic center salem, Tamil Nadu

6.10 Wearables for Blind Population

As mentioned in previous sections, the assistive technology and assistive devices are heavily dependent on wearable systems for regular, rehabilitative, and other clinical applications. The targeted population with all kinds of disabilities in vision uses many varieties of wearables to make improvements in their quality of life. The most significant work done in the area of assistive technology for blind showcases a lot of amazing devices. The reason for developing more wearable devices, apart from regular spectacles and lenses for the people with vision difficulties and blindness, draws a very serious attention. Providing rehabilitation for blind involves a lot of Government's effort, and a wide variety of specialists are involved. Experts like teachers with special trainings, Braille teachers, psychologists, orientation, and mobility specialists, low-vision specialists, and vision rehabilitation therapists are involved in developing assistive wearables for them. Apparently, these efforts and involvements incur a very huge budget.

A sample wearable assistive device is suggested for blinds and is explained in chapter 4 (Section 2).

7

Security Technology and WT

Wearing a mask and hand gloves is not only a well-known practice against the Covid-19, which originated in the Wuhan, China, outburst, but it has also been seen for a long time in medical field while dealing with patients of different ailments during surgery, dressing, and treatments. The wearables that secure medical and caregiving personnel against infectious diseases also come under protection and personal security category, while the security technology commonly refers to the secured financial transactions and information security.

Similarly, wearable head cap or electrode straps on forehead help to make a drowsy driver to receive alarm and bring him back to safe driving mode. Many researchers developed such systems using EEG, EMG, and EOG under various research projects. According to the statement by World Health Organization, more than 1.25 million people are killed in road accidents and around 50 million people are injured in a year, and 90% of such casualties occur in developing countries. As per the statistics of 2017, nearly 1,306 drivers who were involved in fatal crashes (or almost 3%) were reported as being drowsy that causes the accident. These crashes could have been avoided if a proper awakening alarm would have been given to the drivers at the time of drowsiness by a proper monitoring system.

Humans are tending to get drowsiness due to tiredness, boredom, and fatigue. The drowsiness is a state of transition between alert and sleep, leading to a prolonged reaction time as well as a gradual inclination towards sleep. The creation of drowsiness monitoring system is found necessary for the motor vehicle manufacturers and it has become a major focus for safety standard equipment. Drowsiness is also influenced by the mental state, medication, and consumed food of the drivers. When concerning the mental state of drivers, the investigations of the human thinking pattern remain difficult; the plentiful and substantial results obtained in the field of cognitive neuroscience have recently provided an opportunity to solve this problem.

Many wearable systems that intend to detect drowsiness using EEG/ EMG/EOG signals have started coming into practice along with its eye vision-based variants. In EEG-based systems, the noises, eye artifacts (that cause EOG responses), and muscle contraction are removed from the raw signal by appropriate signal filters (pre-processing). The filtered signal is then segregated into EEG sub-bands namely Delta, Alpha, Theta, Beta, and

DOI: 10.1201/9781003052906-7

Gamma based on the corresponding frequency range. According to research, drowsiness results better in Alpha and Theta bands. The linear and nonlinear features of brain states are extracted from both these bands. Now many machine learning algorithms have been developed to find these features and are helpful in decision making to high accuracy rate in detecting drowsiness and switch on the alarm.

7.1 Signature Verification

The need for signature identification system is felt essential for casual to highly secured financial applications for automation. When it comes to highly secured signature verification, it is essential to eliminate forgery attempts by a robust system component, which the image processing-based systems failed to identify with the signatures by skilled forgers. Hence, a more advanced system is necessary to identify the forgers and differentiate between forged signatures and the genuine signatures of the account holders.

7.1.1 Introduction to Online Signature Verification

A work done for eliminating forgery in signature verification by means of a wearable electrode glove by researchers brought in lot of appreciation in the year 2007.

In the old image processing method, a strong impression of the signature is recorded using a prescribed ink on a specified medium, and it is used as the key component for further processing. The image of the signature is then scanned into digital representation and is used for key construction. The process of selecting the match is by searching the database for the nearest match, and the result determines the authentication decision. The image-based signature verification systems use signature images as the only key component; it is possible to access these images for any misuse like forging. Apart from that, the image processing involves huge data, and any reduction in the data volume leads to a compromise with its efficiency. The continuous enhancement to this scenario leads to improving two major criteria in the system. One is to protect the signature from public view and the second is to reduce the volume of data involved so that the speed of the process can be improved. Using the data glove in this new system is a paradigm shift from imaging to signal processing so that the first problem of the old system is immediately solved since there is no significance given to ink or paper. However, the data glove used for this purpose, which the subject wears while signing for authentication or access, consists of many electrodes in various positions of the hand. These electrodes produce continuous signals during the signing process, resemble the image processing in data volume, and need to be reduced to overcome the drawback. PCA which is a popular technique used for source separation has been introduced in this difficult case for both noise and volume reduction. The PCA is found to be competent in solving the problem by the improved results.

As usual, a data glove with all its supporting setup was used in the experiment; three different versions of data were collected, namely, reference, original, and forgery from 50 subjects in this experiment. The average age of subjects involved in this experiment is 37.4 years. There were 30 average reference signatures collected for test data from 30 subjects signing 10 times each to derive the average. The 'original' captioned data is collected from the same 30 subjects signing five times against their respective reference signatures. 'Forged' captioned data is collected from 50 subjects trying to forge the 30 average reference signatures for 5 times each. The subjects from the forging group are allowed to familiarize and practice the target signatures with an unlimited number of trials for forging. The data is recorded once they are confident about forging the authentic reference signatures. The test data includes the training data as a subset of the whole data collected. In each signature data sample $x(i, j)$, if the value of 'j' is high, the dimension of the matrix also becomes larger, which increases the computing load and processing time. In fact, all the data points from a signature are not necessary to identify the key components involved. But at the same time, it is difficult to declare where these key components are present during the whole signing process.

An important problem noted in this setup is the noise factor. Every raw signal from the electrode x is a mixture of the signal of interest 'S' and the background noise 'N' that are present in it due to all minor extra movements and variance in equipment stability: $x = S + N$.

The noises that contaminate the signal of interest vary from time to time, from place to place, and from equipment to equipment, which may affect the genuineness of the signal and, in turn, the system.

The noise removal is done using PCA by selection of Principal Components (PCs) which were then used as the key dataset to represent the voluminous signature data and stand for the Euclidian test in distinguishing the forged signature from the original one. The same selected PCs are used in the reconstruction of the noise-free data signal to confirm the clarity. The signature verification is done when the average reference signatures of subjects are matched with five individual authentic signatures of the same subjects by means of calculating their intra-Euclidian distance. Considering the same average reference, pursuing the 250 forged signatures from the forging group per reference was also carried out. The Euclidian distance for every forged signature $F(f1, f2, f3,..., fn)$ with the reference signature

R(r1, r2, r3,..., rn) is calculated. The noise-free signals extracted using PCA reduce the chance of forged signals getting into acclamation by precisely distinguishing the originals with the distinct Euclidean distance.

The average values of Euclidean distance between the mean reference signatures for the 150 authentic trials are calculated for every dataset from 14- and 5-sensor channels. Following the similar way, the average Euclidean distance for 50 forging subjects against the 30 authentic signatures is calculated. The number of forged signatures considered here is 7050. Evaluation of results shows that riddance of forged signatures from the authentic signature can be easily identified using our PCA-based approach. The False Acceptance Rate (FAR) and False Rejection Rate (FRR) are calculated for the normalized threshold values ranging from 0 to 1. FAR and FRR are calculated by the following formulas:

$$FAR = \frac{\text{Total number of accepted forgeries}}{\text{Total number of accepted forgeries}} \times 100$$

$$FRR = \frac{\text{Total number of rejected genuine}}{\text{Total number of tested genuine}} \times 100$$

The comparative EERs for both 14 and 5 channels are shown in Table 7.1. From the experimental results, we have achieved the EERs of about 3.1% and 7.5% for 14- and 5-electrode data, respectively, with the thresholds of 0.083 and 0.020.

Here, a real-time technique for the easy recognition of handwritten signature is demonstrated. The technique is based on linearly projecting the signature space of data glove into a low-dimensional and noise-free space through the use of PCA. The resulting projections maximize the total scatter across all classes, that is, across all signals of all signatures, and result in a much simpler and efficient approach for signature recognition and verification. This work may be extended to further increase the credibility by involving an artificial neural network classifier.

TABLE 7.1

The comparative equal error rates

Type of channel	Equal error rate (%)
14 channels	3.1
5 channels	7.5

7.2 Augmented/Robust Signature Verification

A novel system suggested by the author and his colleagues resulted in a robust online signature verification system. The novelty lies on the reinforcement to the signals obtained from the data glove-based dynamic signature verification system. This is done by using the photometric measurement values collected simultaneously from photoplethysmography (PPG) along the signing process.

Figure 7.1 shows the augmented online signature verification process as a combination of signals from data glove and the PPG at the same time.

Skilled forgers worldwide try to attempt and succeed in repeating the genuine signatures in many numbers of practices. But the reinforcement suggested using PPG reinforcement brought in a wide gap in the calculation of Euclidian distances between forgers and the genuine template features. This prohibits them from successful forging of genuine signatures which they used to do earlier. This has been proved by the repeated experiments on various subjects using the above combinational features. In addition, the intratrial features captured during the forge attempts also differ widely in the case of forgers and are not consistent with that of a genuine signature. This is caused by the pulse characteristics and degree of bilateral hand

FIGURE 7.1
Online signature capturing along with PPG

dimensional similarity, and the degrees of pulse delay. Since this economical and simple optical-based technology is offering an improved biometric security, it is essential to look for other reinforcements such as the variability factor considerations which we proved of worth considering.

Enhancements to the signature verification systems have been suggested by many researchers [Plamondan, Srihari, S. Rahee, B.H. Yang, H.H. Asada, Y.Y. Gu, Y. zhang, and Y.T. Zhang] and biosignal-based security features are also considered as a unique alternative for applications that require some document evidence like signing checks and other security documents. The signal-based biometrics is the only applicable means for people with physical disability [Andrews Samraj and Nasir Noma]. Making use of multimodal biometric technology, which provides unique and robust identification features for every individual, is in great demand in security environments that require high-quality authentication methods. Using PPG wave forms to distinguish individuals using their biometric component was suggested by researchers in 2007 [J.C. Yao, X.D. Sun, and Y.B. Wan] and is employed in protected applications like e-transactions and access control mechanisms. As a fortification to the current signal-based dynamic signature verification system, we have used a new method by using the combination of the plethysmographic component along with the data glove signals to make the authentication process more robust and distinctive. The opportunity of skilled forging is reduced by the PPG feature that brings in the hand and heart dimensions of an individual into the signature feature vector. In order to further reinforce the efficiency of the system, here it is considered the additional consideration of the intratrial variation approach to further validate the signature process. This method assures the elimination of skilled forging by a multi-level confirmation.

7.2.1 The Modifications in Equipment Setup

The photoplethysmogram (PPG) is a simple equipment that functions on the intensity of light reflected from the skin's surface. The red blood cell count below the skin is considered to determine the volume of blood in that particular area. The recorded signal possesses the measurement of changes in venous blood volume and the arterial blood pulsation in the arterioles, hence representing the heart rate. There are two values supplied by the system and these are the measurements of transmission and reflectance. A sample signal produced by the PPG is shown in Figure 7.2.

Similarly, the same data glove, which we discussed in previous experiments, is used for dynamic signature verification since it is easy to use, free from image and material of signature medium as well as no scanning processes is required. It involves only a direct acquisition of signals from the subjects while they write down their signatures, preprocess it, extract

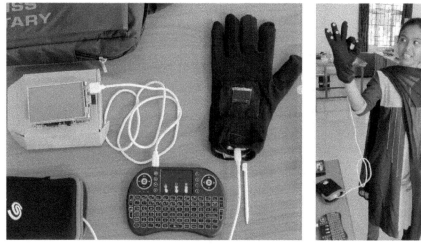

(a) (b)

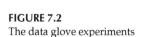

FIGURE 7.2
The data glove experiments

the feature, and match it for classification and decision-making. The data glove offers the users comfort, ease of application, and it comes with a small form factor with multiple application drivers, high data quality, low cross-correlation and high-frequency data lodging. It measures finger flexure (two sensors per finger) as well as the abduction between fingers. The system interfaces with the computer via a cable to the USB port (Platform Independent). It features an auto calibration function, 8-bit flexure and abduction resolution, extreme comfort, low drift, and an open architecture. It can also be operated wirelessly to interface with the computer via Bluetooth technology up to 20 m distance. One glove fits many hands since it is made up of stretchable material 'A'. The data glove and the signature verification process using the glove is shown in Figure 7.2. The output of the probe is fed into the serial port of a pulse oximetry module (from Dolphin Medical, Inc.) Measurements were taken for 50 signatures from 14 sensors of the data glove and four LEDs of plethysmogram fixed on the subject as seen in Figure 7.3.

7.2.2 Subjects and Signal Acquisition Methods

Two subjects were considered as original signers and other four were the skilled forgers. The data glove signals and the peripheral volume pulses (PPG) were sampled at 61 Hz. Both the signals were recorded from the subjects simultaneously while they were signing. The subjects were selected

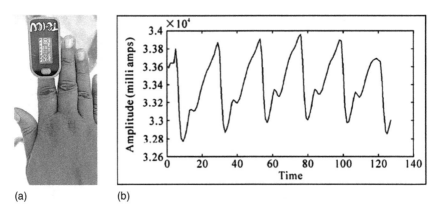

(a)   (b)

FIGURE 7.3
The plethysmogram

among our researchers and the average age of the subjects was 34. The dynamic features of the data glove signal comprise

1) Distinctive patterns to an individual's signature,
2) The hand dimensions,
3) Time taken to complete a signature process,
4) Hand trajectory dependent rolling.

These factors contribute to the feature generated from wearable technology as signal captured from the data glove and make it more suitable to trust for use in signature verification since it provides data on the dynamics of hand movement along the pen movement and the individual's hand dimensions. Along with these four components that represent a person's identity, the heart rate reflected by the wearable plethysmographic signals is also measured as the fifth component to reinforce the system's distinctiveness. The photometric signal consists of the volume of blood that flows through the blood vessels per pulse during every beat of the heart.

7.2.3 Experimental Setup

The recordings of signals were arranged in such a way that one of the subjects writing original signatures was allowed to sign 50 original (own) signatures and the subjects who were assigned to forge were allowed to observe it to do the skilled forgery [B. Majhi, Santhosh, and D. Prasanna Babu]. The forgers were given generous exercises of practice to forge against the two genuine signatures by giving special sessions to practice the original signatures.

The subjects were seated in a comfortable chair located in a sound-protected room. The data glove was fixed on their right hand and the photoelectric probe was fixed to the index finger of their left hand. All the subjects appointed for forging did sign 50 forge signatures one by one each with the help of a tracing paper placed on the original signatures after successful training. The subjects were asked to write the signatures, in two sessions, with an interval of 24 hours. Fifty signatures were collected per subject in one session. The skilled forging of the original signatures from forging subjects were also collected in the same intervals. Forging with 50 signatures per original subject took a total of 100 signatures per session for two original signatures. The PPG signal during the signing in process was also recorded for every subject from all the 14 electrodes embedded in the data glove. Hence, there were 200 original signatures and 800 forgings were recorded and considered for analysis. Similarly, 1000 simultaneous PPG recordings were also included in the analysis.

7.2.4 Preprocessing and Feature Vector Construction

The dimension of each recording of hand glove signal, A, is of order n by m, where n is the number of electrodes and m is the number of sequential samples per second. n is fixed as 14 throughout the experiment, and m differs in milliseconds as the intra- and intersubject vary in signature timings.

The dimension of every PPG, matrix B is j by k, where j is the number of LEDs and k is the number of sequential samples in one second. Throughout the experiment, j was fixed as 4 and k was taken up to the exact time length of m. To condense the dimension and to reduce the effects of overlapping spectral information between noise and signature features, Singular Value Decomposition (SVD) approach was applied to both matrices A and B. Since there is a real factorization for any real nX m matrices, it is determined that the SVD protects the characteristics of the source signal matrix given by the m signal samplings from n electrodes. Matrix B used to incorporate PPG representation was also subjected to exactly similar SVD process to estimate the singular values for use in feature vector. The average size of the glove signature matrix is (14,234) as well as the average size of the PPG signal matrix is (4,234). After the application of SVD the features are reduced to (14,1) and (4,1), respectively. We were used the l-largest singular values of A as well as q-largest singular values of B as feature contents representing every single data glove signal and PPG respectively. Therefore, the entire signal A is now represented by a highly discriminate feature vector of length A (l) and the entire PPG is represented by B (q). These l and q largest singular value features of A and B contain the feature component of the subjects' unique signature ID that discriminates the original from forge signatures. To minimize computational complexity, we set the l value to be five and q to 2 throughout these experiments.

7.2.5 Matching and Classification

The reference signature along with the reference PPG FG (Genuine Factor) was computed from a set of reference enrollment samples. The pair having minimal overall angle to the rest of signature, PPG pair, was selected as the reference signature to which all the comparisons were carried out. The genuineness of any factor pair Fi A. SAMRAJ ET AL. Copyright © 2010 SciRes. JIS 27 is decided by the similarity factor (SF) and both the components of FG & Fi are calculated as the angle between their principle subspaces.

7.2.6 Results and Discussions

Euclidean distance between the genuine reference signature PPG fusion factors with other genuine signature and forge signatures with the corresponding PPG of the subjects are calculated. Ten random sample distances across the two sessions were shown taken for considerations and other signatures were also giving similar results. These results were reported in our previous works. The Equal Error Rate (EER) can be calculated if and only if a set of FAR and FRR are available. In this experiment, both are found to be zero and hence could not able to draw a curve of FAR and FRR to find the intersection point which is EER. As an enhancement to this system we intended to find the consistency of the signatures written by the forgers with that of the consistency of the genuine signatories. This is to identify the best forger and later this factor may be used to enhance the authenticity of the entire system using the distinct quality of intertrial coherence. Table 7.1 shows the results in terms of Euclidian distance between the signatures cum PPG fusion template to every subject's with their own signature cum PPG features found in different trials. We found that the consistency of the genuine signatory is consistence and all the other four skilled forgers were not able to retain their consistency across the trails. This can be seen from the zigzag lines. The performance of the data glove declines with the reduction of sensors. The performance degradation with the reduction of electrode channels from the hand glove was reported in previous works to minimize the hardware and volume overheads in data processing. But the proposed technique of this work helps to eliminate the said problem by providing strong reinforcements provided in two levels. The first one being the PPG factors and the second one is the intratrial variability.

Table 7.2 shows the values of intratrial variability with the corresponding reference temples of every subject. In all the 10 trials, the Euclidean distance is calculated against the templates of the individual subjects' training to write the same signature. The signature and PPG combination is not altered throughout the experiment.

This wide gap between the intratrial variability reveals that this factor can be considered as an improvement factor in reinforcing the robustness of data glove + PPG-based signature verification system.

TABLE 7.2

Intratrial variations in Euclidian distance between individual templates to corresponding individual signatures

	Trial1	Trial2	Trial3	Trial4	Trial5	Trial6	Trial7	Trial8	Trial9	Trial10	Average
Genuine	34370	9934	10759	15392	21582	15866	11819	875.38	27423	19917	16793
Forger1	165410	53161	198290	191950	215300	28236	218790	83738	118500	201180	147455
Forger2	44223	51702	12488	8906.1	33876	54862	21046	15533	12128	6096.4	26086
Forger3	46886	127180	103760	123760	165250	46779	75789	129740	62206	124990	100634
Forger4	82633	82805	249120	62179	73780	68638	138800	81469	49637	184130	107319

7.2.7 Conclusions

The proposed intratrial variability measurement of multimodal signature + PPG-based signature verification system is found to be reliable in strengthening the identification of genuine subjects of data glove based signature verification system. The novelty lies in two levels of using PPG factors and its augment to the robustness of the data glove features as well as the counting of intratrial variability factors against the reference signatures. In feature, the data glove may be fabricated as additionally accommodating the PPG sensor to make it easy for everyone. This technique is verified with the present easy modeling of data glove signals using SVD. Two sets of singular vectors produced by SVD are fused as the feature, where the primary set is from data glove with the maximum energy of the signature during the process of signature writing, and the secondary set is derived from the photometric signals extracted by PPG simultaneously during the process. These selected set of vectors are known as the principle subspace of data glove output matrix A and PPG output matrix B, respectively. These principle subspace sets are used to model a reinforced signature feature robust against any forgery attack. This research work is a venture to demonstrate that the intratrial variability factor enhances the signature verification system that uses the PPG as its combination. This novel system is much potential to offer a sensitive high-level security for applications like banking, electronic commerce, and legal proceedings than the existing similar systems. On the other hand this founding strongly supports the reduction in hardware by manufacture data glove integrated with PPG with minimum sensors so that the size of the equipment can be made simple and efficient for handling by a single hand. Since there is the possibility of reducing the feature size by means of reducing sensor in the data glove as well as reducing timing in the PPG, we can achieve a low-cost signature verification system suitable to a common user in common place.

7.3 Emergency Response System for Elderly/Disabled/ Persons in Ambulance

The design of the emergency response wearable system was made to match the closest situations of the usage by the elderly and disabled or ambulatory subjects (people/person). The materials chosen are a sponge, an iron material, and a coin. The objects are chosen such that they are available in conventional daily-life situations to closely resemble the user environments in future. Also to show the contrast in the texture of the objects that may cause varieties of responses while using to produce simple holds with the data gloves. Data from five healthy volunteer subjects, including two female subjects, have been taken into consideration. The average age for the subjects is 19.5 years. The reason for involving healthy subjects is to avoid troubling elderly with much number of repeated trials and tiresome procedures, test runs, and iterative functions. These intensive tests are essential to find out the best possible and optimally suitable paradigm for the intended users.

The volunteers were highly cooperative with an intention to help this noble cause with a clear understanding of the benefits of the projected system. The glove is worn in the right hand since it is the dominant hand of all five subjects and the data are transmitted through the wireless transmission.

An electrodes-embedded wearable data glove, as depicted, is used to capture the hand movements of five subjects. The hand movements were categorized by capturing the signals from the glove while the subjects were asked to soft hold and hard hold a firm shape (coin) and a tool(like Pen) for special movement like rotation. In the first phase, six paradigms were designed and in the second phase two more paradigms were added for testing. Both the phases were carried for all five subjects.

In phase – 1, P1.1 Coin with Light holds (CL), P1.2 Coin with Tight holds (CT), and in phase – 2, P2.1 Rotation of a Coin (CR), P2.2 Rotation of a Pen (PR)

7.3.1 Data Description

The data were recorded from the data glove and were stored as separate files for each trial of gesture for the given paradigm. The data stored in a single file has 14 columns as 14 electrodes captured the gesture actions and the rows of each sample matrix range between 265 or 266. The variation is due to the time difference that happens in storing the files when a gesture is done. The row size is uniformly taken as 265 for all processes to avoid disparity. For all the subjects, these ten trials of the feature extraction for each category of gesture paradigms were done. Therefore, the total number of files collected was 400 (5X8 X10= 400). Each of these 400 files has 14 columns and 256 rows which come to 3584 data points per file. So, for five subjects, the data points come to 1,433,600 in numbers. The captured data signals are voluminous in nature due to the sampling rate, the use of 14-electrode glove and number of trials. In this research work, a coin rotation and pen rotation movements were taken into consideration for detailed analysis. The dataset had been collected from the 'Centre for Ubiquitous Computing and Communication (CUCC)' of Multimedia University. The research center is working in the areas of wearable computing and biometrics. MATLAB software has been used for obtaining the results in the association between paradigms, intratrial variability and optimization of the paradigms in this research work.

a. IRON MATERIAL TIGHT HOLD (HT)
   The iron material is tightly held in this paradigm with the data glove worn by the subjects. While holding the iron material tightly, the

pressure given by the subjects is high compared to the tight hold of the sponge. All the 14 sensors are now affected by this paradigm and could be observed by the data matrix of the extracted signals. The bend of the fingers is considered as maximum in this paradigm compared to the light hold of the sponge.

b. COIN LIGHT HOLD (CL)

The reason for choosing a coin (any flat material) is to show the difference in the objects by the pressure experienced by the two fingers of the subjects. The usage of coin is trial with certain precaution as it does not slip away due to its tiny nature. The paradigm of lightly holding a coin with the data glove worn hand is challenging due to the less weight of the coin and also the light hold.

While holding the coin, the bend sensors in all the fingers were undergoing more change in their bends than the rest of the materials as sponge or iron materials since the size of the item drastically varies. The bend of fingers is considered as maximum in this paradigm comparatively like the light hold of the sponge material and iron material. All the five subjects carried out this gesture with attention.

c. COIN TIGHT HOLD (CT)

The coin is tightly held with the data glove worn by the subjects. In this gesture paradigm of holding the coin tightly, the pressure given by the subjects is considerably higher to the light hold of the coin. All the 14 sensors are contributing to the corresponding changes in the data matrix of the extracted signals along with the thumb and index finger with maximum contribution. The bend of the thumb is considered more when compared to the thumb of any other paradigms used in this work.

d. COIN ROTATION (CR) and PEN ROTATION (PR)

The rotation of the coin and the pen is used for the peer study of the comparison in paradigms. These were undergone to check out the difference in the data matrix of the thumb and the index fingers in case of the reduction of the electrodes and also to find the change in data when flipping is done.

7.3.2 Wearable Emergency Response System

This system helps to identify the effective gestures for communication between caretakers and patients. The optimization of the system has been made in three parts – identifying the overlapping paradigms which infer the diversified paradigms (association between paradigms), exercising the similar gestures with ambiguity in resemblance (intratrial variability of paradigms), and minimizing the quantity of electrodes and the complexities of input (optimization in quantity of electrodes). The results of the three parts are helpful in setting the efficient response protocol for the emergency-response communication system. The biosignals acquired by the data gloves are explained in the next sections of this chapter. The description of the data gloves from which the data signals were acquired for the system is clearly given. The paradigms for the glove-based communication system are explained with its nature and the way the signal captured from those paradigms is pictured. The advantages in the proposed schemes are Error free-wearable Infrastructure, Multi channel and energy efficient device that reduce human task and enhanced cognitive control.

7.3.3 Feature Extraction Techniques for System Implementation

Feature extraction is the process of simplifying the data representation into an accurate reduced dimension while extracting along with the associated characteristics for the preferred responsibilities (Chu Kiong Loo et al. 2011). The feature extraction techniques are chosen with a careful analysis to satisfy three criterions. Firstly, the volume of data has to be addressed and properly represented by the feature extraction techniques. Secondly, the representation should be sharp enough to identify the variations of different paradigms and gestures even shown by the same person. And thirdly, the complexity is to be under minimum levels and the time taken for generating the feature should be minimal. The classification of the features depends upon the efficacy of the extracted features. If the classification is carried out without a thriving feature, extraction on a superfluous and high-dimensional data would be computationally complex. Satisfying the above three critical characteristics for this specific application, the feature extraction techniques like SVD, Fractal Dimension (FD), Fast Fourier Transforms (FFT) and SVD with Average Distribution (AD) were taken for appliance.

7.3.3.1 Singular Value Decomposition

A data glove of m number of electrodes (sensors) each generating n number of samples with respect to time is taken into consideration and produces an output matrix A m x n. The matrix A represents the feature contents and the SVD has been calculated (Shohel Sayeed et al. 2006; Shohel Sayeed et al. 2009; Chuanjun Li et al. 2006; Chuanjun Li et al. 2007). It has been found in many signal processing applications and control systems that the SVD of matrix formed from observed data can be used to improve methods of signal parameter estimation and system identification (Nidal S. Kamel et al. 2008).

7.3.3.2 Fractal Dimension (FD)

FD is a statistical measure to indicate the complexity of an object over some region of space or time interval. This method of feature extraction has been widely used in various domains. FD has various estimation methods. For better classification results, the most appropriate method of FD should be used. The various FD methods are Katz's method, Higuchi's method, and Rescaled range method, Time-dependent FD, Differential FD, Differential signals. The general term for factor characterization is FD. The complications of similar figures are categorized by the FD. This creates a pattern with the basic building blocks of the signals (B. S. Raghavendra and D. Narayana Dutt 2010).

7.3.3.3 KATZ'S Method

The FD of the sample is done using the Katz's method of estimating the FD (Umut Guclu et al. 2011). The sum and average of the Euclidean distance between the successive points of the sample (L – sum of Euclidean distance between successive points of samples and average of Euclidean distance between successive points of samples) are calculated as well as the maximum distance between the first point and any other point of sample (d).

7.3.3.4 Fast Fourier Transformations (FFT)

The FFT is used to approximate the captured data glove signals and to plot the support vectors. Here, radix-2 FFT algorithm employed for the first two elements is taken for butterfly model, based on the stage of computation. The K value varies from 0, 1, 2, 3, and N is the total number of available data. The FFT value is calculated using MATLAB function.

7.3.3.5 SVD with Average Distribution

The projected technique introduced to enhance the SVD-based feature selection process is the AD. This is carried out by getting the average of readings from each channel and then finding out the maximum and minimum average values. Their corresponding channels are identified. Uniform random sample values from only these two selected channels are collected. This is done as a statistical method of uniform distribution. The SVD technique is applied to the values obtained by the AD technique which helps to enhance the feature selection process. This helps in reducing the number of electrodes by adopting the electrodes producing maximum and minimum values as sole responsible for representing the entire feature matrix. Hence, the electrodes producing maximum and minimum values are chosen as random value producers, and in a constant sampling interval (Down Sampling) the data points are selected from a complete trial. All these signals are selected from the two diversified electrodes. The feature vectors from all the trials of a subject are taken in to account for averaging to form a single feature representing a particular gesture of that subject.

7.3.4 Classification Techniques

7.3.4.1 Euclidean Distance

Euclidean distance is used to measure the distance between each pair of singular values.

7.3.4.2 Linear Discriminant Analysis

LDA is used to classify objects into groups depending on the set of features taken. Here, multi-class LDA is used to classify multiple unknown data glove signals into multiple classes (Mohammad Shahin Mahanta and Konstantinos N. Plataniotis 2012).

7.3.5 Association of Paradigms

The gesture paradigms designed for the 'Association between Paradigms' are the daily use objects such as a sponge, an iron material, a coin, and a pen. The reason for selecting multiple paradigms is to enable its future utilization in automated systems with multiple choices. The paradigms are not always meant to satisfy a particular need like the need of water, calling the care givers, the change of position, the medicine needed, etc. The protocols are set such that the system can be designed according to the need of the elderly or disabled on a case-to-case basis. This facilitates the systems to benefit many users with many paradigms assigned for many protocols. The signals are acquired from an electrodes-embedded wearable data glove which is used to capture the hand movements of five subjects.

The hand movements are categorized by capturing the signals from the glove while the subjects are allowed to soft hold and hard hold (a) a soft material (sponge), (b) a hard material (iron), and (c) a coin. The 10 samples of each hold were captured under six paradigms as sponge light hold (SL), sponge hard hold (SH), iron material light hold (HL), iron material tight hold (HT), coin light hold (CL), coin tight hold (CT), coin rotation (CR), and pen rotation (PR) for all five subjects. The row size of each sample matrix is either 265 or 266, and the column size is 14. The row size is uniformly taken as 265. Ten trials of the feature extraction for each category are done. The data signals are voluminous in nature due to the sampling rate and the use of 14-electrode glove.

7.3.5.1 Association Between Paradigms

Using SVD with Euclidean distance reduces the dimensionality of the data signals and estimating the feature vector. The SVD values of all samples of five subjects are calculated and the graphs are plotted, respectively. The classification is done by the distance-based classification technique using the Euclidean distance. In this estimation method the Euclidean distance is calculated among the thresholds and features for linear classification of classes and found the results were similar. To make it specific the SVD-based Euclidean distance is taken for further analysis.

7.3.5.2 Results of the Experiments

The SVD features are calculated for each sample for each paradigm. This is continued for all five subjects. The SVD values are calculated and tabulated for ten samples of five subjects. In the feature vector, the first 4 values give significant differentiation to various paradigms as the remaining values produces very low difference. The Euclidean distance is used to measure the distance between each pair of singular values. Here, Euclidean distance is calculated for the SVD features of reference signal and the SVD features of other gestures.

The average SVD values are calculated for the same subject.
SVD VALUES OF LIGHT HOLD OF SPONGE (SL) FOR SUBJECT – 3
Samples and First 4 SVD features are as follows:

SL1 116730, 0.0114, 0.0082, 0.004

SL2 115110, 0.0657, 0.0214, 0.0083

SL3 114920, 0.0199, 0.0071, 0.0047

SL4 115000, 0.0286, 0.0083, 0.0047

SL5 114720, 0.0666, 0.0303, 0.0087

SL6 115590, 0.0302, 0.0089, 0.0052

SL7 115970, 0.0254, 0.0091, 0.0054

SL8 115060, 0.0661, 0.0228, 0.0074

SL9 116520, 0.0204, 0.017, 0.0076

SL10 115490, 0.0657, 0.0243, 0.0112

Average 115510, 0.04, 0.0157, 0.0067

The specific reason for this research is, to avoid any misinterpretation of gestures as another in the paradigm. The recommendations for selecting candidate gestures for an effective emergency communication interface is given based on the findings. The wide variety of gestures with their Euclidean distances from the reference gesture signal is done for Subject – 1 to Subject – 5 for all 10 trials per subject and found similar in all cases.

7.3.5.3 Discussions

The SVD features of single user are calculated for all six paradigms and the results are produced. The average SVD is calculated for all the five subjects under all six paradigms and individually represented in graphs.

The produced results are compared in various combinations with all subjects interrelatively. The holding of coin tightly exhibit a distinct pattern from holding a soft material lightly. As a contra, holding hard material tightly is similar to holding coin light and falls under similar distance from the reference.

7.3.5.4 Recommendations Based on the Experiment

Holding soft material lightly can be combined with holding soft material tightly for good difference.

Holding soft material lightly can be combined with holding hard material tightly or holding coin lightly for better difference.

Holding soft material lightly can be combined with holding coin tightly for the maximum difference.

The combination of holding coin light and holding hard material tight should be avoided since they overlap in Euclidean distance with the reference pattern.

7.3.6 Intratrial Variability

The intratrial variability is carried out to find the gesture fluency of a subject with respect to the paradigms. A gesture fluency of a subject is the easiest gesture movements the subject can perform at any point of time without difficulty and not much variation. The average difference in measurements of a particular gesture output on various trials (repetitions) by a same subject gives the intratrial variation. The 10 samples of signals from each paradigms like – sponge light hold, sponge hard hold, iron material light hold, iron material tight hold, coin light hold, and coin tight hold, respectively, denoted as SL, SH, HL, HT, CL, CT. Here also the size of each sample matrix for its row falls between 265 and 266. By conducting multiple trials of all subjects, the intratrial variability of paradigms provides results to prove that the selections of paradigms for use in recommended systems are widely set apart and are safe from misinterpretation. These results suggest the paradigms that are to be and not to be used in glove-based emergency communication systems. The sample graphs for the intratrial variability are given for clear knowledge. A 0% correlation with no overlaps in FFT features, classified by LDA for third subject's reference with 10th trial signal captured from soft material hard-hold gesture action, is found.

7.3.6.1 Discussions and Recommendations Based on ITV

The use of intratrial variability is to find the most intact gesture which is proved with the best classification results. The order of the paradigms is listed from most intact-able to least intact-able as CT, HT,

CL, HL, SL, and SH. The highest overlap is found in CT for all five subjects with 93.5% to 91.5%. The lowest overlap is found in SH for all five subjects with 30% to 40%. These intraclassification distances are analyzed and are found intact with minimal distances. Even though they are closer distances than the intergesture distances, the overlaps alone are accounted as best intact points. Table 4.3 consists of the FAR, FRR, and EER of various combinations in percentage (%) values by using Naïve Bayes Classifier (Shohel Sayeed 2007a). The feature extraction was continued by the FFT for classification.

It is interestingly seen from the average EER results that the 'Light Holds' are prone to much overlaps than the 'Tight Holds' as in the case of CT and SH. As a controversial result, HL possesses less EER than HT, and naturally it is due to the reason that the hard material cannot be pressed or squeezed more as in the case of light material and a coin. This explains the very good reasoning of arranging the paradigms with the hard materials or uses the hard hold of other materials in system design. The light hold of soft materials can be complemented with the hard hold or hard materials to reduce the error rate. The FFT features resulted in better realizations while classifying them, which helped us in identifying the best distinguishable paradigms from each other that provided better results. The minimum FAR is found in the HT from HL paradigm as 0.83% and minimum FRR 0.67% is found in the HT to SL and the minimum ERR is in HT from HL as 0.415%. Similarly, the maximum FAR is found in SL from HT paradigm as 29.82% and the maximum FRR in HL as 34.69% and maximum EER in CT as 19.13% paradigm. Therefore, the recommendations of the paradigms that can be combined for the projected user interface should be like SH, HL, and CT and should not be like SL, HT, and CL.

7.3.6.2 *Optimization of This Work*

The paradigms are divided into two groups, namely hold group and rotation group. In the emergency response communication, the pen and coin are used as the materials and the materials are given an activity of rotation. The data glove used for acquiring the signals is a 3D input device; hence, it is suitable for a wide range of applications like control and manipulation of virtual worlds, gesture and cognitive media, physiotherapy rehabilitation, and control device for artists in remote controlled environments, so the addition of rotation is tested. The optimization is done with five volunteered subjects. Each subject repeats every paradigm gesture 10 times and it is recorded separately. By calculating the average of all 10 trials, the intrareference is formed for the particular gesture signal of the subject involved. The comparison of intraclassification distance is the distance of reference signal against individual signals. These intraclassification distances are analyzed and are found intact with minimal distances. Even though they are closer distances than the intergesture distances, the overlaps alone are accounted as best intact points. From the signals, the SVD with AD features are calculated. The paradigms taken for the process are Coin Rotation (CR) and Pen Rotation (PR). The row size of each sample matrix is either 265 or 266, and the column size is 14. The row size is uniformly fixed as 265.

7.3.6.3 Singular Value Decomposition with Random Average Distribution

The novel technique introduced in optimization of electrodes is such that the selection of electrodes was made by choosing two electrodes with the major difference between them. So in general, the electrode producing highest bend value and the electrode producing the lowest bend value produced by the bend sensors will be chosen. This provides a very clear identification of the highly involved channel of the particular gesture and the non-contributing channel of the same. The data signals acquired are from 14-electrode glove and optimization is done to reduce the dimensionality of the data while estimating the feature vector. The signals of every subject for every trial from 14 electrodes are considered as the matrix for the random process. The electrodes producing the maximum value and the minimum value are chosen as the random variables in a constant sampling interval. The thumb near and ring little are the two electrode positions with the maximum difference in electrode values and are taken for the difference of these random variables at constant interval. All these difference values contribute to the feature vectors and represent the complete signal from the two electrodes. The feature vectors from all the trials are subjected to averaging to form a single feature representing a particular gesture by a subject. The expected average value of a signal trial is calculated using a simple method in the following way. The SVD with AD of every signal trial is calculated and used.

1. The signals of every subject for every trial from 14 electrodes are considered as the matrix for the random process.
2. The electrodes producing the maximum value and the minimum value are chosen as the random variables in constant sampling intervals.
3. The differences of these random variables at constant intervals are calculated.
4. All these difference values contribute to the feature vectors and represent the complete signal from the two electrodes.
5. The feature vectors from all the trials are subjected to averaging to form a single feature representing a particular gesture by a subject.

7.3.6.4 Contribution of the Research Work

The novelty could be seen in the work starting from the utilization of the signals acquired from the data glove for the purpose of gesture recognition, feature construction, optimization of feature vector, selection rationale of paradigms, optimization by cost, and complexity etc.

- A novel application for the data glove was projected, tested for feasibility, implemented, optimized, and presented to the disabled community.
- The SVD, FD, and FFT tools for feature extractions were applied on the acquired signals to test their fitness and efficiency of response.
- The best choice of paradigms and gestures were found out with iterative trials to avoid overlaps and to ensure the maximum diversification among gesture features.
- The close observation of intratrial variability for a particular gesture of individual subjects helps to find the gesture fluency of every individual subject. This gesture fluency and useful finding of the investigations would surely help customizing the glove-based emergency response communication system to exactly suit different users.
- The optimization done by the reduction of the number of electrodes, dimension reduction, and improvement in the deflection of features makes the system more professional.

8

Strategic Operation Technology and WT

Strategic Operation Technology is a predefined and ambiguity-free application which has its structure and operational plan well defined, and no dangling references or uncontrolled flow happens. The system functions are definite and zero fault occurrence is assured at any state of situation. This chapter deals with such an effort to bring some of the Wearable Technology applications with well-defined operation route map to fall under this category to make use of them in very critical operation for which reliability is the basic requirement.

Mostly space, surveillance, rescue, lifesaving and computer vision projects need such definite and zero fault systems. Projects like autonomous cars need to know the direction of pedestrian movements such as left, right, forward, or backward. In pedestrian movements, events like fall or sudden change in direction due to running in opposite directions caused by dilemma need to be alerted as exceptions by the automated driving system at the earliest to make decisions to apply break or negotiating a direction change. Instead of such image-based applications, that involve cognitive approaches, movement prediction approaches, etc. a perfect signal-based environment setting, on a predefined movement setup is under the trend of research now.

The Strategic Operation Technology is a customized system developed rapidly with reliable building blocks and ever failing interfaces to respond to any situation that requires a hi-tech intervention.

8.1 The Need for Simple and Fast Feature Identification

The aim of such simple and fast feature identification work is to find a more reasonable and less complex feature extraction technique to make it convenient and simple for the classification algorithm to differentiate.

A gesture-based wearable technology application with the wide thresholds, eliminating less active or less contributing input channels to reduce computational resources in this gesture-based application, helps to understand such a hardware setup to optimize the operational data.

This experiment was done with a wearable glove that produces continuous and variable signals from all its electrodes with respect to the degree of

DOI: 10.1201/9781003052906-8

flex being generated by every gesture. The signals differ according to the movement of hand, the movement of fingers, the duration of the gesture activities, the degree of bending of fingers, the speed of completing the action, and the hand dimension of every single user using the glove. The cause of continuous refinement to this work is to enhance the life of elderly and disabled by providing more accurate and versatile system for them to communicate and control machines for their assistance. As a secondary benefit, the perfect and faulty gestures can be identified for the use or elimination from the setup, respectively. The data acquisition was done through the glove and represented in 8-bit flexure resolution at the given sampling rate.

The experiment was done among 10 volunteers, and they were asked to repeat the experiment for 6 instances. The data was collected using 14 columns and the row was fixed to 250. The objects used for the experiment were a book, a stone, a mobile, and a bottle.

8.1.1 Modeling the Signal by Zone

The simplification of the entire work of modeling was done in few steps.

1. Four best active channels out of 14 channels were selected for each trail of data (the channels may vary from trail to trail, gesture to gesture and subject to subject).

   S is the selected channel set $S = \{ c_p, c_q, c_r, c_s \}$

where

$$c_p, c_q, c_r \,\&, c_s \ are\ the\ subset\ of$$

$$C = \{C_1, C_2, C_3, \ldots C_{14}\}$$

2. The method of selecting the four channels is by the standard deviation value (the higher the flex duration the higher is the standard deviation).

3. The selected four channels (S) with the highest ranking of the standard deviation for every trail are further reduced to two channels by voting method from the table as the best active channels for this gesture by the subject. (This is done by voting mechanism and is explained in Table 8.1 for the gesture 'lifting of mobile' by four subjects).

4. Every single signal from each selected channel for a particular gesture was divided into five zones based on latency. The maximum and minimum value of the particular signal in every zone for every trail is identified and presented in Table 8.3.

TABLE 8.1

Selection of channels for subjects by voting

Subject 1 – Lifting Mobile				Selection Procedure	Subject 3 – Lifting Mobile				Selection Procedure
Trail 1	E	J	M	L		Trail 1	D	I	K	L	
Trail 2	M	E	G	I		Trail 2	D	L	K	H	
Trail 3	E	L	M	G	M -Ring / Little & E – Index Near	Trail 3	D	L	K	C	D-Index Near & K-Ring Far
Trail 4	L	M	J	E		Trail 4	D	I	K	L	
Trail 5	M	E	I	J		Trail 5	D	I	K	L	
Trail 6	E	I	J	M		Trail 6	D	I	K	N	

Subject 2 – Lifting Mobile				Selection Procedure	Subject 4- Mobile Lifting				Selection Procedure
Trail 1	D	G	I	K		Trail 1	D	F	K	L	
Trail 2	D	F	G	K		Trail 2	D	F	K	L	D-Index Near & L-Ring / Little
Trail 3	A	D	G	K	D-Index Near & G-Middle Near	Trail 3	D	F	G	L	
Trail 4	A	D	G	K		Trail 4	D	G	K	L	
Trail 5	A	D	G	K		Trail 5	D	F	K	L	
Trail 6	D	F	G	K		Trail 6	D	G	K	L	

8.1.2 Standard Deviation as a Comparison Feature

The standard deviation is calculated by

$$\sigma = \sqrt{\frac{1}{N} \sum_{i=1}^{N} (x - \bar{x})^2}$$

where x is the mean of the channel output, which is calculated as
AVERAGE(number$_1$,number$_2$,...number$_n$.)
and n is the sample size. The standard deviation is calculated to identify the most active channels [14]. This is done for reducing feature size, computational complexity by eliminating low-performing channels and reduction of hardware.

8.1.3 Results of Comparison Features

Each signal data is divided into five different time zones and the maximum and minimum values for every time zone are noted from these time zones. This is to identify the activity pattern of the particular subject over different segments of time.

Table 8.2 is developed as the highest and lowest parts of every signal, S1,.... S6 from channels M and D across the time zones. This denotes the high and low instances of the actions performed. Indirectly this value helps to ensure

TABLE 8.2

Standard deviations of maximum and minimum values from ring/little and thumb/index for all trails

Subject 1 – Lifting Mobile	Values/Sig	S1	S2	S3	S4	S5	S6
Ring/	Max	63.31114	65.02538	69.45646	70.10492	62.8156	15.92168
Little	Min	21.61481	20.22869	11.84483	52.27619	51.40331	6.708204
Thumb/	Max	70.91192	43.87254	61.24541	61.18415	67.56996	80.67961
Index	Min	15.16575	35.08133	48.01042	47.88737	49.26256	21.25324

TABLE 8.3

Modeling of Lifting of Book Gesture by SUBJECT 1

Subject 1 – lifting book	Zone 1	S1	S2	S3	S4	S5	S6
M	Max	3144	3135	3168	3148	3135	3169
	Min	3135	3131	3139	3135	3103	3168
E	Max	1803	1863	1919	1945	1890	1934
	Min	1770	1811	1789	1800	1744	1779
	Zone 2	**S1**	**S2**	**S3**	**S4**	**S5**	**S6**
M	Max	3151	3131	3166	3145	3135	3168
	Min	3143	3121	3158	3139	3130	3168
E	Max	1774	1811	1787	1797	1752	1775
	Min	1770	1774	1777	1784	1746	1761
	Zone 3	**S1**	**S2**	**S3**	**S4**	**S5**	**S6**
M	Max	3163	3168	3164	3164	3178	3168
	Min	3151	3121	3151	3135	3130	3168
E	Max	1778	1851	1797	1850	1843	1767
	Min	1770	1773	1775	1782	1745	1762
	Zone 4	**S1**	**S2**	**S3**	**S4**	**S5**	**S6**
M	Max	3172	3248	3169	3257	3231	3175
	Min	3165	3167	3166	3166	3184	3168
	Max	1905	1925	1893	1926	1915	1807
E	Min	1780	1858	1806	1856	1848	1767
	Min	1805	1793	1683	1722	1724	1812

the uniformity among the signals across all trails. The standard deviation for all these times zones for their minimum and maximum is calculated and listed in Table 8.4. This is to again identify the pattern more clearly.

Similarly the calculations are done for data from other objects as bottle, mobile, stone and the standard deviation tables are given below (Tables 8.1, 8.2, 8.3, 8.4, and 8.5) (Figures 8.1 and 8.2).

TABLE 8.4

Zonewise standard deviation for Lifting of Book from All Trails of subject 1

Zone 1			Zone 2			Zone 3			Zone 4			Zone 5	
M	Max	15.32	M	Max	15.4	M	Max	5.57	M	Max	41.06	M	Max	40.08
	Min	20.71		Min	17.4		Min	17.14		Min	7.25		Min	39.89
E	Max	53.04	E	Max	20.5	E	Max	38.20	E	Max	44.96	E	Max	11.13
	Min	23.71		Min	13.4		Min	12.95		Min	40.31		Min	53.69

TABLE 8.5

Comparison with previous methods [6] and [7]

Method 1	Method 2	Proposed technique
SVD (Singular Value Decomposition)	Fractal Dimension & SVD	Simple raw data from Selective channels
Euclidean Distance	Euclidean Distance	Standard Deviation

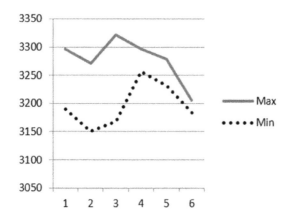

FIGURE 8.1
Zone 5 of M

8.1.4 Promises Found in the Wearable Strategic Operation Technology

The modeling is done by the simple analysis of checking whether the active signal falls in between the upper and lower limits of the signal model created for that particular gesture by the particular subject during the training sessions.

The condition is that the active signal should be within the boundaries for all the five zones so that without confusion the active signal can be adjudged as the member of given gesture model.

The more the training signals are taken for modeling, the boundaries becomes robust for the particular gesture.

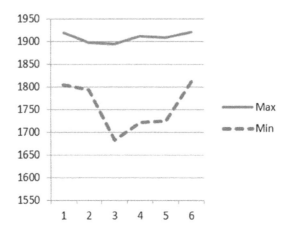

FIGURE 8.2
Zone 5 of E

These boundaries differ from subject to subject even for the same gesture. The reason is because the hand dimensions, the pace of performing the gesture, and the style are contributing to these differences. This has been seen in the previous chapters as a biometric augmentation.

8.2 Consequences of Strategic Operational Technology Using Wearable Technology

The understanding of activity, signals produced by the activity, and the characteristics of the signal over time zones, active electrodes, and consistency in patterns are the subject of interest in this research work. The proposed work clearly distinguishes these entire pattern-making activities and helps to reduce the feature size with a vigilance parameter. If the maximum and minimum values within time zones act as the feature values for a particular activity by a specific subject, the standard deviation acts as the vigilance parameter for every time zone.

This experimental work of finding features by simplest technique involves no mathematical modeling or functions but simply identifies the activity from the best possible channels and distinguishes the activity in different time horizons clearly. It helps in reducing the operational data and makes it convenient for the applications associated with this to produce or interpret effective communication for the disabled.

This experimental work boosts the process of gesture translation by eliminating the construction of complex features and eliminating the time overheads. The reduction in dimension from the raw data acquired through the data glove is made simple with no compromising of quality and precision.

8.3 Fast and Easy Gesture Recognition by Wearable in Strategic Operations

The idea started from a health-care monitoring system. In the health-care monitoring on a clinical base involves many implicit communications between the patient and the caretakers. Any misinterpretation leads to adverse effects. A simple wearable system can precisely interpret the implicit communication to the caretakers or to an automated support device. Simple and obvious hand movements can be used for the above purpose. The following experiment with multiple gestures is executed to form a novel methodology simpler than the existing sign language interpretations for such implicit communication. You can see how the experimental results show a well-distinguished realization of different hand movement activities using a wearable sensor medium and the interpretation results always show significant thresholds.

The demand for caretakers for elderly and the disabled has increased in great ratio. At the same time the population of the elderly in hospitals and care homes has also started increasing. The patients with Parkinson diseases definitely need a caretaker to do all their daily activities. The high cost involved in this service of expertise makes it always an unachievable target for the health-care organizations. To overcome these issues of caregivers, the robotic assistance would be the precise and appropriate solution. Hence, this following experiment is presented to understand how a wearable helps the people in need to achieve this target (Figure 8.3).

An electrodes-embedded wearable data glove is used to capture the hand movements of an elderly, patient, or an injured subject who is in treatment. The aim of this experiment is to enhance the utility of precise communication with the caretaker, including communicating for accessing robotic assistance in modern health care, by conveying the message by gesture made by the patients or elderly as symbolic. The usage of such digital conversions from the hand movements to message passing can go beyond human comprehension and mere communication interfaces. Hence, an effective and clear message passing would be ideal to have robotic assistance. However, controlling robots, a technologically advanced device, poses complex interfaces and is not user-friendly. In order to make the process simpler, this system incorporates a small biosensor embedded glove which captures the dimensions of the fingers and their movements. This system can overcome the limitations of human's aid due to tiredness and lack of timely service. This can also

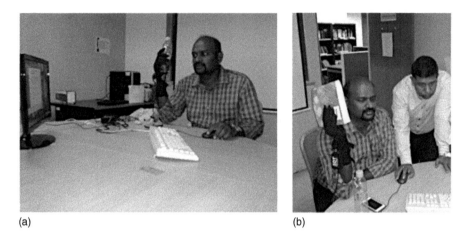

(a)   (b)

FIGURE 8.3
Wearable communication experiment using Emergency Gesture by patients and caretakers of elderly and disabled

break the barrier of the elderly and disabled toward operating the robotic system for assistance. The glove is worn to any convenient hand of the subjects and the data are transmitted through the wireless transmission.

8.3.1 Mode of Experiments

The interaction between the human and computer technologies increasingly and continuously provides natural ways to operate and communicate with machines. Ranging from speech to vision, all the standalone to wearable interaction technologies help to change the way how people operate computers. With all these interaction methods, gesture recognition through wearable technology takes an important and unique role in human communication with machines.

The familiar data glove, a device used in similar experiments was used here also on hands to facilitate the process of sensing and quantifies the fine-motion control in this hybrid system with robotics. It is a very important and new dimension in the field of medicine and health care at this point of time since epidemic diseases may require robotic medics to fight against them in saving human life. Most input devices that can be fixed on a patient offer limited degrees of freedom; whereas the data glove is unique in that it offers multiple degrees of freedom for each finger and hand as well. This flexibility facilitates the user to have seamless communication with the computer to greater extent than most other input devices (Figure 8.4).

In this experimental work, a sponge, an iron material and a coin – the objects available in conventional daily life environments were used as contact elements of the subjects using it. This assorted selection is to show the

Sensor	Description
0	Thumb flexure (lower joint)
1	Thumb flexure (second joint)
2	Thumb-index finger abduction
3	Index finger flexure (at knuckle)
4	Index finger flexure (second joint)
5	Index-middle finger abduction
6	Middle finger flexure (at knuckle)
7	Middle-ring finger (second joint)
8	Middle-ring finger abduction
9	Ring finger flexure (at knuckle)
10	Ring finger flexure (second joint)
11	Ring-little finger abduction
12	Little finger flexure (at knuckle)
13	Little finger flexure (second joint)

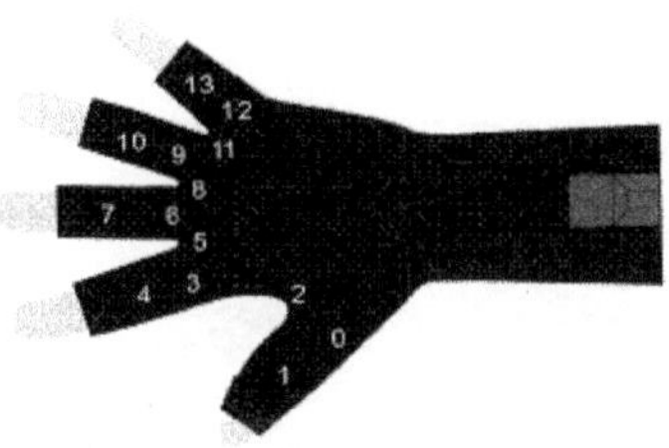

FIGURE 8.4
Position of the sensors embedded in hand glove

contrast in the texture of the objects that may cause varieties of responses while using to produce simple holds using data gloves.

Data from five healthy subjects, including two female subjects, have been taken into consideration, whose average age is 19.5 years. The glove is worn in the right hand since it is the dominant hand of all five subjects (Figure 8.5).

In all experimental trials of this paradigm, the same electrodes-embedded wearable data glove is used to capture the hand movements of five subjects. The hand movements are categorized by capturing the signals from the glove while the subjects are allowed to soft hold and hard hold (a) a soft material (sponge), (b) a hard material (iron), and (c) a coin. Ten samples of each hold were captured under six paradigms as sponge light hold (SL), sponge hard hold (SH), iron material light hold (HL), iron material tight hold (HT), coin light hold (CL), and coin tight hold (CT) for all five subjects. The row size of each sample matrix is either 265 or 266, and the column size is 14. The row size is uniformly taken as 265. The feature extraction is done using two methods from all the captured signal samples by Singular Value

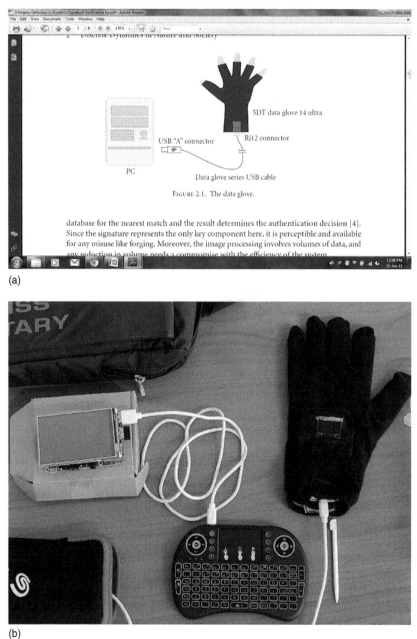

(a)

(b)

FIGURE 8.5
A wearable data glove connected to the system

Decomposition (SVD) and Fractal Dimension (FD). The data signals are voluminous in nature due to the sampling rate and the use of 14-electrode glove; hence, to reduce the dimensionality of the data signals while estimating the feature vector, SVD and FD are being used. In both estimation methods, the Euclidean distance is calculated among the set thresholds and features for linear classification of classes. The thresholds were selected based on the experimental simulation trials.

8.3.2 Use of Singular Value Decomposition (SVD) to Quantify Feature Set

The output of the data glove of m (number of electrodes) sensors each generating n number of samples with respect to time is taken into consideration and produced as matrix A_{mxn}. The matrix A represents the feature contents and the Singular Value Decomposition has been calculated as

$$A = U_{mxn}.S_{mxn}.V_{nxn}^{T} \qquad (8.1)$$

where U and V are real orthogonal matrices and S is a real pseudo diagonal matrix with non-negative diagonal element.

It has been found in many signal processing applications and control systems that the Singular Value Decomposition of matrix formed from observed data can be used to improve methods of signal parameter estimation and system identification.

8.3.3 Use of Fractal Dimension (FD) to Quantify the Feature Set

The Fractal Dimension of the sample is done using the Katz's method of estimating the fractal dimension.

The sum and average of the Euclidean distance between the successive points of the sample (L – sum of Euclidean distance between successive points of samples and a – average of Euclidean distance between successive points of samples) are calculated as well as the maximum distance between the first point and any other point of sample (d). The fractal dimension of the sample (D) then becomes

$$D = \frac{\log\left(\frac{L}{a}\right)}{\log\left(\frac{d}{a}\right)} \qquad (8.2)$$

$$= \frac{\log(n)}{\log(n) + \log\left(\frac{d}{L}\right)}$$

wheren is L divided by a.

8.3.4 Results of the Experiment

The SVD features are calculated and tabulated for 10 samples of 5 subjects. In the feature vector, the first 'n' (few) values give significant differentiation to various paradigms. So the figures provided here are represented for 'n' feature values (Tables 8.6, 8.7, 8.8, 8.9, 8.10, 8.11) (Figures 8.6, 8.7, 8.8).

The fractal dimension is calculated from 10 samples of all 5 subjects. The variation in the fractal dimension of each category is given in bar charts.

The fractal dimension is calculated when $fs = 35$, for six paradigms as sponge with light hold (SL), hard hold (SH); iron material with light hold (HL), tight hold (HT); coin with light hold (CL), tight hold (CT) (Figures 8.9, 8.10, 8.11).

The graphs represent the average SVD values of each subject and then the average of all five subjects (Figures 8.12, 8.13, 8.14).

TABLE 8.6

SVD values of sponge with light hold (SL) of sub-3

sl1	1.0e+005 *	1.1673	0.0114	0.0082	0.004
sl2	1.0e+005 *	1.1511	0.0657	0.0214	0.0083
sl3	1.0e+005 *	1.1492	0.0199	0.0071	0.0047
sl4	1.0e+005 *	1.15	0.0286	0.0083	0.0047
sl5	1.0e+005 *	1.1472	0.0666	0.0303	0.0087
sl6	1.0e+005 *	1.1559	0.0302	0.0089	0.0052
sl7	1.0e+005 *	1.1597	0.0254	0.0091	0.0054
sl8	1.0e+005 *	1.1506	0.0661	0.0228	0.0074
sl9	1.0e+005 *	1.1652	0.0204	0.017	0.0076
sl10	1.0e+005 *	1.1549	0.0657	0.0243	0.0112
Average	1.0e+005 *	1.1551	0.04	0.0157	0.0067

TABLE 8.7

SVD values of sponge with hard hold (SH) of sub-3

sh1	1.0e+005 *	1.1767	0.037	0.0178	0.0122
sh2	1.0e+005 *	1.214	0.0427	0.0211	0.0111
sh3	1.0e+005 *	1.1977	0.0407	0.0258	0.0128
sh4	1.0e+005 *	1.1869	0.0666	0.0371	0.0171
sh5	1.0e+005 *	1.1956	0.0666	0.043	0.0163
sh6	1.0e+005 *	1.1797	0.0677	0.0401	0.0289
sh7	1.0e+005 *	1.1981	0.068	0.0478	0.0235
sh8	1.0e+005 *	1.1971	0.0678	0.0387	0.0231
sh9	1.0e+005 *	1.1952	0.0683	0.0412	0.0233
sh10	1.0e+005 *	1.2128	0.0518	0.0247	0.0137
Average	1.0e+005 *	1.1954	0.0577	0.0337	0.0188

TABLE 8.8

SVD values of iron material with light hold (HL) of sub-3

hl1	1.0e+005 *	1.1822	0.0675	0.0231	0.0207
hl2	1.0e+005 *	1.2113	0.0714	0.0361	0.0165
hl3	1.0e+005 *	1.2214	0.0696	0.0178	0.0118
hl4	1.0e+005 *	1.2121	0.0691	0.0184	0.0152
hl5	1.0e+005 *	1.205	0.02	0.0178	0.0074
hl6	1.0e+005 *	1.2025	0.0682	0.0187	0.0141
hl8	1.0e+005 *	1.2085	0.0687	0.0195	0.0149
hl9	1.0e+005 *	1.2052	0.0684	0.0339	0.016
hl10	1.0e+005 *	1.2023	0.0319	0.0171	0.0138
Average	1.0e+005 *	1.2056	0.0594	0.0224	0.0144

TABLE 8.9

SVD values of iron material with tight hold (HT) of sub-3

ht1	1.0e+005 *	1.2213	0.0691	0.0524	0.0207
ht2	1.0e+005 *	1.2359	0.0798	0.0175	0.014
ht3	1.0e+005 *	1.2306	0.0798	0.0648	0.0186
ht4	1.0e+005 *	1.2368	0.0714	0.0175	0.0141
ht5	1.0e+005 *	1.2268	0.0725	0.064	0.0185
ht6	1.0e+005 *	1.2111	0.0547	0.0214	0.0157
ht7	1.0e+005 *	1.223	0.0715	0.0599	0.0166
ht8	1.0e+005 *	1.2251	0.071	0.0617	0.0171
ht9	1.0e+005 *	1.2312	0.0761	0.0626	0.0159
ht10	1.0e+005 *	1.2256	0.0731	0.0591	0.0166
Average	1.0e+005 *	1.22674	0.0719	0.04809	0.01678

TABLE 8.10

SVD values of coin with light hold (CL) of sub-3

cl1	1.0e+005 *	1.275	0.054	0.0255	0.0188
cl2	1.0e+005 *	1.2533	0.0418	0.019	0.0182
cl3	1.0e+005 *	1.2692	0.0455	0.0229	0.0162
cl4	1.0e+005 *	1.243	0.0683	0.0402	0.0195
cl5	1.0e+005 *	1.2624	0.0702	0.0446	0.0252
cl6	1.0e+005 *	1.2668	0.0472	0.0204	0.0151
cl7	1.0e+005 *	1.2472	0.0399	0.019	0.0166
cl8	1.0e+005 *	1.2505	0.0435	0.0242	0.0178
cl9	1.0e+005 *	1.258	0.0688	0.05	0.0209
cl10	1.0e+005 *	1.2565	0.0684	0.0476	0.0219
Average	1.0e+005 *	1.2581	0.0547	0.0313	0.0190

TABLE 8.11

SVD values of coin with tight hold (CT) of sub-3

ct1	1.0e+005 *	1.2892	0.0593	0.0313	0.0248
ct2	1.0e+005 *	1.2903	0.0714	0.06	0.0338
ct3	1.0e+005 *	1.2984	0.0731	0.0617	0.0348
ct4	1.0e+005 *	1.2786	0.0697	0.0574	0.0299
ct5	1.0e+005 *	1.2851	0.0703	0.0539	0.0284
ct6	1.0e+005 *	1.3102	0.0582	0.0314	0.0205
ct7	1.0e+005 *	1.2758	0.0676	0.057	0.0296
ct8	1.0e+005 *	1.2959	0.056	0.0346	0.0172
ct9	1.0e+005 *	1.2857	0.0677	0.0553	0.0284
ct10	1.0e+005 *	1.29	0.0733	0.0548	0.0335
Average	1.0e+005 *	1.28992	0.06666	0.04974	0.0281

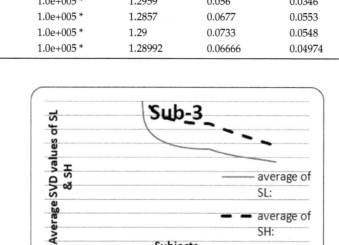

FIGURE 8.6
Comparison of average SVD values of SL and SH of sub-3

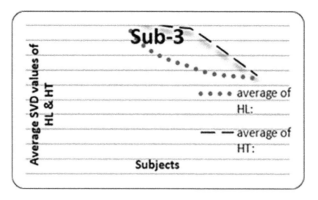

FIGURE 8.7
Comparison of average SVD values of HL and HT of sub-3

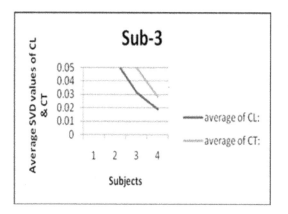

FIGURE 8.8
Comparison of average SVD values of CL and CT of sub-3

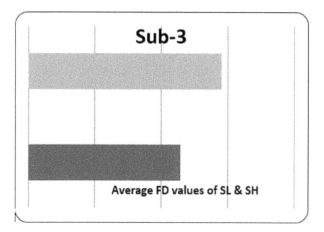

FIGURE 8.9
Comparison of average FD values of SL and SH of sub-3

The average of FD features of each subject is calculated and then the average of all five subjects is taken and graphs are drawn (Figures 8.15, 8.16, 8.17).

The results are compared in various combinations and selected results are stated below (Figures 8.18, 8.19, 8.20, 8.21, 8.22, 8.23, 8.24, 8.25).

Euclidean distance is used to measure the distance between each pair of singular values. In general, the distance between points x and y in Euclidean space is given as

$$D = \sqrt{(x-a)^2 + (y-b)^2}$$

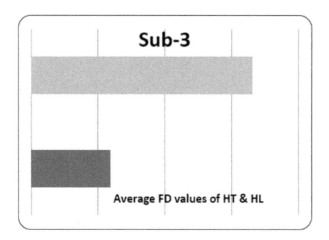

FIGURE 8.10
Comparison of average FD values of HL and HT of sub-3

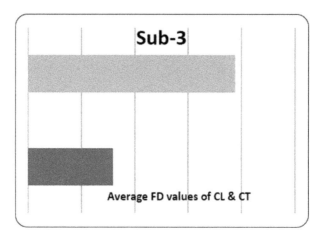

FIGURE 8.11
Comparison of average FD values of CL and CT of sub-3

Here, Euclidean distance is calculated for the reference signal and the SVD values using the formula (Figure 8.26),

$$d = sqrt\left[(x)^2 + (y)^2\right]$$

The SVD and FD features of all samples of five subjects are calculated and the graphs are plotted, respectively. The classification is done by the distance-based classification technique using the Euclidean distance.

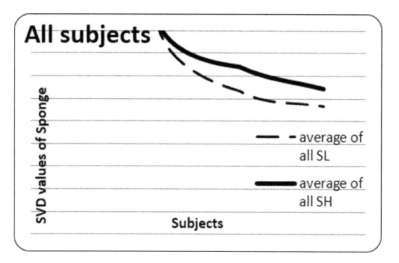

FIGURE 8.12
Average of SVD values of sponge of all five subjects

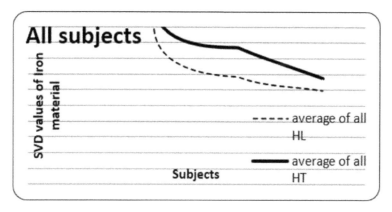

FIGURE 8.13
Average of SVD values of iron material of all five subjects

8.3.5 Discussion on the Experiment and Its Results

The American Sign Language (ASL) recognition system developed by many researchers has been the only remedy for emergency communications. But the proposed work simplifies the total framework by making minimal efforts free from any formal language semantics.

A cyber glove is used in the systems and the movements are traced by a Flock of Birds 3-D motion tracker to extract the gesture features. The usage

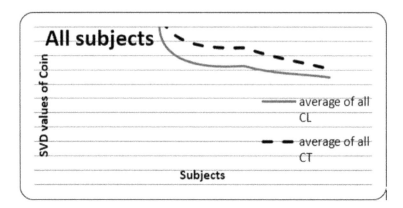

FIGURE 8.14
Average of SVD values of coin of all five subjects

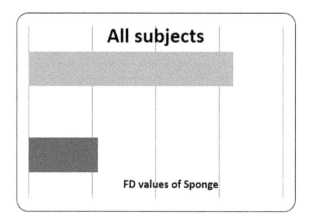

FIGURE 8.15
Average of FD values of sponge of all five subjects

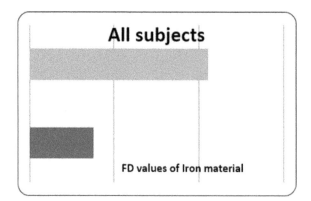

FIGURE 8.16
Average of FD values of iron material of all five subjects

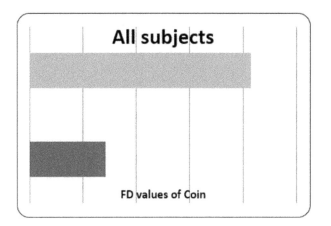

FIGURE 8.17
Average of FD values of coin of all five subjects

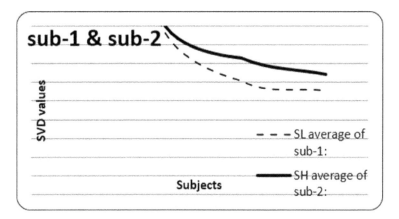

FIGURE 8.18
Average SVD values of sponge light hold (SL) and sponge hard hold (SH)

of these motion trackers is expensive and ASL should be taught prior to the elderly and the disabled which is difficult. Moreover, such functions are difficult during unbearable pain and emergency.

The SVD and FD features of single user (say sub-3) are calculated for all six paradigms and the results are produced. The average SVD and FD are calculated for all the five subjects under all six paradigms and individually represented in graphs. For in-depth study, the produced results are compared in various combinations with all subjects interrelatively.

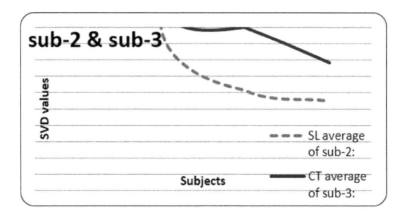

FIGURE 8.19
Average SVD values of sponge light hold (SL) and coin tight hold (CT)

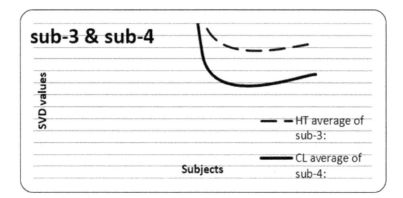

FIGURE 8.20
Average SVD values of iron material tight hold (HT) and coin light hold (CL)

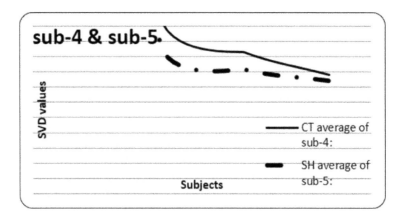

FIGURE 8.21
Average SVD values of coin tight hold (CT) and sponge hard hold (SH)

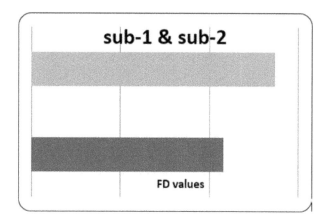

FIGURE 8.22
Average FD values of sponge light hold (SL) and sponge hard hold (SH)

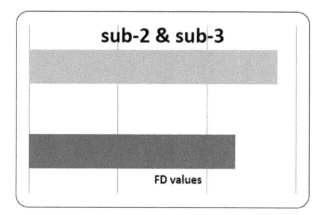

FIGURE 8.23
Average FD values of sponge light hold (SL) and coin tight hold (CT)

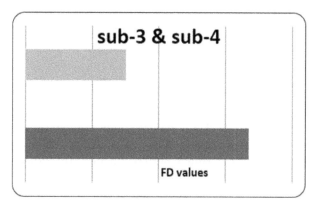

FIGURE 8.24
Average FD values of coin light hold (CL) and iron material tight hold (HT)

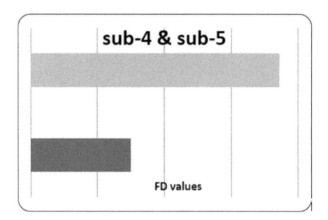

FIGURE 8.25
Average FD values of coin tight hold (CT) and sponge hard hold (SH)

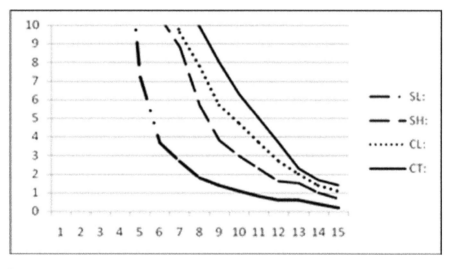

FIGURE 8.26
Intra-Euclidean distance of SVD features of subject-5 on various paradigms

8.3.6 Consolidation of Gesture Movement

The affective gesture movements suggested in this example work for the interaction of the elderly and the disabled with caretakers are found to be a successful way of communication by the obtained experimental results. The proposed method is free from complex functions providing simple natural gestures for the people to adopt. Hence, a wearable data glove is much useful for the emergency communication of the people under challenged conditions as it captures the signals that are generated by the mere finger and hand movements. The results obtained by the proposed system show the significant variation of the signals when different objects are held in different applied pressure. The system can be enhanced as a people-friendly, moderate cost and easy accessible robotic control system in health care, replacing mankind.

The experiments using simple gestures which are seen so far has to be reinforced for an error-free interpretation. For which a simple and obvious hand movements may be used for preliminary testing experiments. Whereas for the complete product the challenge in interpretation rest on the clear and well-discrete input classes extracted from the wearables. The later reinforcement experiments suggest well-distinct classes of gestures that are suitable for such system developments. It further presents a clear understanding about the groupings that are not suitable for such implied communications. The experiment resulted in the realization of suggested hand movement activity groups over the coupling of other movements for forming classes. These findings helped in establishing extended and significant thresholds between gestures which are desired and reliable for fault-free communication.

In detail, the fine tuning and quick interpretation of gesture-based communication between caretakers and elderly or the disabled is mandatory for every error-free emergency response communication session or a system. The motive for confirming it to be faultless is as vital as that of taking a decision on selecting a drug for injection to the patient. Any wrong understanding of gestures by the life support systems attached to them may turn deadly. The nonstop improvement of the prevalent gesture-based communication system requires a strong and wide boundary of thresholds to discriminate each response from other. The high speed and wabbly hand movement in this process makes it always an unattainable target for the systems while using overlapping gesture signs for communication. To augment the vibrant command interpretation, to support the partial or fully automated robotic assistance or communication system, scheming of effective gestures by selecting extremely distinct activity to produce highly differentiable signals are necessary. Hence, in many of the experiments conducted for the system development, with the help of an electrodes embedded wearable data glove to capture the different hand movements of a subject is investigated for suitability. The aim in conducting such experiments is to identify most distinct activities that can be performed by the hand gestures to differentiate symbolic representation such as 'feeling good' and 'feeling pain'. While using the life support system operated by the hand glove produced gesture inputs the wide thresholds provide a good safety zone for the safety of the patients using it. The usage of such digital conversions from the hand gestures can go beyond human communication interfaces by involving robotic assistance that can be operated on the effective classification of gestures. However, controlling robots, a technologically advanced device, poses a great demand of clarity in given commands since there involves complex interfaces and is not user friendly. The experiments were carried out using a system that consists of a wearable glove which captures the gestures shown using patient's finger and hand movements. The reason for explaining this work is to break the barrier of the elderly and disabled toward operating the robotic system or machines for assistance. Actually, the identification of right and wrong combinations of gesture clusters in this work helped to improve the system design with larger threshold possibilities.

This reinforcement experiment was conducted with the electrodes-embedded wearable data glove as depicted in the previous experiments to capture the hand movements from five subjects. The hand movements are categorized by taking the signals from the glove while the subjects are allowed to soft hold and hard hold (a) a soft material like sponge, (b) a hard material like wood or iron bar, and (c) a coin. Ten samples of each hold were captured under six paradigms as sponge light hold (SL), sponge hard hold (SH), iron material light hold (HL), iron material tight hold (HT), coin light hold (CL), and coin tight hold (CT) for all five subjects. The row size of each sample matrix is either 265 or 266 and the column size is 14. The row size is uniformly taken as 265. Ten trials of the feature extraction for each category are done. The data signals are voluminous in nature due to the sampling rate and the use of 14-electrode glove; hence to reduce the dimensionality of the data signals while estimating the feature vector, two modeling methods of Singular Value Decomposition (SVD) and Fractal Dimension (FD) is used. Even though FD is not taken in to the scope of this work, in both estimation methods the Euclidean distance is calculated among the thresholds and features for linear classification of classes and found the results were similar. To make it specific, the SVD-based Euclidean distance alone is taken for further analysis in this experiment.

The SVD values are calculated and tabulated average taken from 10 samples of 5 subjects. In the feature vector, the first 'n' (selective) values give significant differentiation to various paradigms. So the figures provided here as example in Table 8.12 are represented for 'n = 4' average feature values.

TABLE 8.12

Average feature construction for 5 subjects out of 10 samples in 6 category

	SL	SH	CL	CT	HL	HT
Sub1	115510.1	119540.1	125294.0	128545.0	128206.1	129624.0
Sub2	114133.2	118749.1	122911.1	124981.1	121210.1	122674.1
Sub3	113978.1	117986.0	125810.1	128992.1	124631.1	125679.1
Sub4	118411.3	121019.1	126609.1	129567.1	125310.1	126438.0
Sub5	115432.3	120011.1	124377.1	127471.1	121211.1	123681.0

Euclidean distance is used to measure the distance between each pair of singular values, in general, the distance between points x and y. The SVD values of all samples of five subjects are calculated and the graphs are plotted respectively. The classification is done by the distance-based classification technique using the Euclidean distance. The specific reason for this research is to avoid any misinterpretation of gestures as another in the paradigm. The recommendations for selecting candidate gestures for an effective emergency communication interface is given based on the findings represented in Figure 8.3 to Figure 8.8. Since the figures for all the intratrials and all the subjects reflect similar patterns, only random choices used to explain the outcome due to insufficient space. The holding of coin tightly exhibits a distinct pattern from holding a soft material lightly. As a contra holding hard material tightly is similar to holding coin light and falls under similar distance from the reference. This is done for Subject 1 to 5 for all 10 trials per subject and found similar in all cases. The average Euclidean distance with reference signal of all 10 trials for each subject was calculated and the same nature and distance pattern is found to be continuing. The distinct gesture continues to be the same when the average patterns of Euclidean distance for all five subjects were plotted out.

8.3.7 Recommendations

Holding soft material lightly can be combined with holding soft material tightly for good difference.

- Holding soft material lightly can be combined with holding hard material tightly or holding coin lightly for better difference.
- Holding soft material lightly can be combined with holding coin tightly for the maximum difference.
- The combination of holding coin light and holding hard material tight should be avoided since they overlap in Euclidean distance with the reference pattern.

8.3.8 Discussions

Various gestures developed by the patients during emergency to communicate need proper interpretation. Any wrong prediction leads to adverse treatment. The gesture recognition system developed nowadays needs to be fine tuned and kept error free before implementation of emergency communications. This proposed work not only simplifies the framework but also enhances the credibility by making optimal efforts to form distinct semantics. A cyber glove is used in the systems and the movements are traced by a Flock of Birds 3-D motion tracker to extract the gesture features. The usage of these motion trackers is expensive and ASL should be taught prior to the elderly and the disabled which is difficult. Moreover such functions are difficult during unbearable pain and emergency. The SVD and FD values of single user (say sub-3) are calculated for all six paradigms and the results are produced. The average SVD and FD are calculated for all the five subjects under all six paradigms and individually represented in graphs. For in-depth study, the produced results are compared in various combinations with all subjects interrelatively.

8.3.9 Conclusion of the Experiment

Amidst all the affective gesture movements suggested in this experiment, a set of gestures stand alone distinctly and share not common boundaries with other gestures. Using such distinct gestures for a sensitive system, such as emergency response communication, always benefits the purpose. The proposed gestures are not much complex to implement and are simple and natural for anyone to adopt. Hence, a wearable data glove as a basic device for the emergency communication of the people under challenged conditions as it captures the signals that are generated by the mere finger and hand movements is well enhanced by these recommended selective gestures. The results of such enhancement by the proposed gesture models show the substantial variation of the signals when different objects are held in different applied pressure and activity. The system can be further enhanced as a people friendly, moderate cost, and easily accessible robotic control system in health care, replacing uninterrupted, intensive, and tiresome manual interaction.

8.4 Zero Error Wearable Technology Applications

Wearable technology has already become a part of Space, Defense, and swift strategic operations. Too much dependency on too many support systems becomes complex in these fields of operations. Standalone, and independent support systems are much desired for a safer operation. Especially, the speed of decision making on a streaming input is very important with a less complex but accurate output producing systems. Alternative to powerful systems is less complex system with definite results based on a simple technique helps a lot of time and lifesaving. Hence, the above system discussed in Section 8.2 is an ideal example to deploy in such situations.

9

Software and Power Requirements of WT

The software required to operate the wearable devices are not alien to the coding community. We need not to learn a new programming language or operating system to develop and operate a wearable device. A slight modification and deviation from routines will suffice. Having it in mind that the device that the software is going to run is a wearable and not a conventional desktop is sufficient to create well-aligned codes for the purpose. Mobile operating systems help a lot in connecting to the wearable device since the plug and play mechanism and the USB ports, IoT concept, all came to give hand in developing the software for wearables.

Power is the life of any hardware or software system; graceful degradation is not acceptable in performance of the wearable system and is expected to be its best at any point of time to deliver the desired output. So, power source maintenance and alternatives need to be considered in the design of wearable systems used in crucial applications.

9.1 Writing Codes for Wearable Devices

There is no separate programming language for wearable computing. But a basic hardware knowledge of electronics is very much a desired requirement for all who venture into embedded programming for wearables. Any conventional embedded system coding is sufficient. Usually, the developers of wearables write their codes in high-level languages like C, C++, or Python and translate them into microcodes using appropriate software and download them on the firmware environment, i.e. on a microcontroller chip for executing them along with adding necessary formatting and interfacing. But at the same time Low level programming is also an available choice in complex and more sophisticated embedded programming since the developer have more customized control over the hardware.

The olden days embedded systems used Assembly code for operations. Generally, we know that Assembly Language is the closest layer next to the machine code instructions and produces small size hex files; the problem with assembly is its weak portability and slow processing and development of codes.

DOI: 10.1201/9781003052906-9

So, the language for the present embedded systems is selected due to their efficiency in handling the small memory size and the ability to run the program faster since the hardware should not be slowed down because of its slow running software. As it is well known, the next important thing is the portability. Portability is best when the same program is compiled for many different varieties of processors. The best help of such portability is in easy implementation like plug and play, trouble-free maintenance and readability for correction or modification.

A microcontroller is a microprocessor dealing with restricted and specialized operations instead of a conventional microprocessor which does all general-purpose processing. Choosing an appropriate microcontroller is important to do the microprogramming efficiently. The microcontrollers belong to different families and are sold in a wide variety like 4 bit, 8 bit, 16 bit, and 32 bit processing. 32 bit microcontrollers are the ones used in recent embedded systems. In addition to other skills, the developer should be well versed in basic functions of voltage, power, current, resistance, data sheets of components, etc. for dealing with the boards and circuits.

An advantage is that microcontrollers always come with some standard software tools as well as simulation tools. This support is provided by genuine manufacturers as tool kits that are really helpful. They include assemblers, compilers, debuggers, and a simulating tool. So it is always recommended to buy genuine products from genuine vendors who have a good knowledge in supporting you in both component quality and in software support. Buying other products available and essential for the interfacing with the microcontroller is good for an easy construction of the wearable.

ARM, (7, 9, 11) Arduino, Adafruit, Stitichkit,Cortex-M0..M4, and Linux are the specific software dealing with wearable programming. Their simple library functions help the programmers to make many best utilities available to their codes easily. Many developers create coding for any wearable product that is being created. Mobile app or standalone device coding are done with appropriate software based on the demand and nature. Researchers He jiang, Xin chen et al., in their article 'Software for wearable devices: Challenges and opportunities' have mentioned that still many inconsistencies exist between operating systems and the software that is being used for wearables.

9.2 Platforms for Development

Wearable technology needs two platforms for its hardware and software. They are design platform for hardware and operating system for software. WICED studio (Wireless Internet Connectivity for Embedded Devices) is a development platform for wearables. This platform is capable of connecting to Wi-Fi, TCP/IP, Cloud Communications, Socket-based Communication, and other peripherals.

Positioning technology can be integrated into wearables to widen their scope and applications. WaRP (Wearable Reference Platform), an ARM architecture-based design platform for wearables, enables such flexible, scalable inclusions and works with operating systems like Android and Linux.

Android wear is another SDK platform released by Google for users to download and use. Apps can be developed using this android wear since many tools directly support wearable device development.

Tizen is another open source operating system, which is used by Samsung Galaxy Gear2 watch. Tizen has many tools that are worth to be incorporated for developing wearables.

9.3 Understanding the Components of a Device in Operation

A recent wearable application that uses a digital camera to identify the given food for the vision impaired has a charge-coupled device as its image sensor. Some cameras use CMOS as image sensor for the purpose. An analog to digital convertor and a processor which process the digital data, with some actuators and a memory to store the data, are the components of the digital camera system. Now a days people prefer using Arduino, PIC, or Raspberry pie controllers to their Projects as per the complexity of the requirements.

So it is easy to learn coding to an Arduino and Raspberry Pic and other smaller versions of them once developers are familiar with the coding. A sample embedded c code to display characters on a LCD display starts like this:

```
#include <reg51.h>
#define kam p0
Void lcd_initi();
Void lcd_dat(unsigned char);
Void lcd_cmd(unsigned char);
Void delay();
Void display(unsigned char *s, unsigned char r)
```

Mostly all basic embedded codes are available for free as open source and further modifications could be done if the developers are well versed in the high-level language constructs.

9.4 Power Requirements for Wearables

A robot electronic circuit requires 5v and 10 v DC converted from s 220-230v AC if connected to a power source. Software like Eagle helps to draw the power circuit connection of the robotic component. The same idea is applicable to standalone wearables with larger size. But for the smaller size wearables small 3.7v DC batteries are available with recharge options. Tiny wearables have power sources to a very small size with and without lithium. Wireless charging batteries contribute to a faster and easy recharging of wearable devices.

Pin-type lithium-ion battery is used in various wearable devices. They are less than 25mm in length and 4mm in width and weigh less than or equal to 1 gm only. This helps to expand the style, design options of the wearables. Rapid charging is another helpful technology nowadays to make recharging more convenient. The power discharge also ranges from 30mA to 60 mA depending on the model. Apart from the pin type, we can see flexible card-like batteries that go easily into fabric pockets and are glued with wearables. These thin film batteries can provide the capacity of 1mAh to 1kAh and beyond.

Energy harvesting is another alternative for these small batteries to get recharged while they are depleting in operation.

9.5 Energy Harvesting in Wearables for Self-power

Many futuristic and novel designs of energy harvesting for wearables started emerging due to a good amount of research focus in this area. Many power management circuits are designed for utilizing multiple sources of energy harvesting. This will reduce the frequency of regular recharging and improve the system efficiency.

Harvesting energy from the human body is the best suitable technology for the wearables due to its close proximity with human body. Piezoelectric transducers located on the wrist area of a person can get power supply ($330\mu W\ cm^{-3}$) to recharge the batteries of wearables. According to the direct Piezoelectric effect, the electric displacement is equal to the sum of products of stress tensor with the direct piezo electric charge coefficient and with the product of the electric field and dielectric permittivity.

The other possible avenues from where the energy could be harvested are kinetic energy, light, temperature, and radio frequency. An EMG-based gesture recognizing system that works on a wearable band in wrist can use photovoltaic cells to harvest energy from light. Thermoelectric generators

(TEGs) can power such wearables by converting human body temperature into 280µV. A direct conversion of human motion into electric signals is made possible with piezoelectric material with no other input in any form.

A wearable system can be divided into (1) Energy source, (2) Energy harvesting, (3) Utility core wearable, (4) Application and communication interface. Any communication like Zigbee, Wi-Fi, Bluetooth, or RFID uses more power than the core signal processing or AD conversion, etc. So instead of lithium ion (Li-ion) batteries the Nickal metal hydride (NiMH) should be used due to their high depth of discharge (DOD). The better place for energy harvesting is the body joints since their motion amplitude is higher than other muscles of the body.

Interestingly, the face masks people are wearing to protect them from virus infection or from cold have a new technology of producing power when breathing from mouth can be converted into electric charge by attaching rubber band and stretchable golden electrodes connected to the PTFE films layered in a PDMS structure.

10

Higher-Order Human–Robot Interface

10.1 Human Interface and Robotic Interface

Interfacing a system with humans or with robots is entirely a different concept. The conventional connection of wearable devices with humans is by means of multiple sensors and represents a cyber-physical system. However, it is easy to connect a wearable device to a robot since it is a simple machine-to-machine connection.

So, a triangle connection of man, wearable, and a robotic machine interface is interesting and makes the humans to connect to machines more easily than before. We use different types of human–machine interfaces through wearables. Identification/recognition of people through such wearables and human biometrics or cognition is an interesting way of gaining an authenticated machine control.

Gesture computing is a novel area of computing which deals with identification of people based on their gestures, (gait analysis) identification of gestures to find what is being told by the gesture, and monitoring of treatment given to various types of patients whose limbs are affected by one or more reasons to ensure their return to normal.

A unique combination of wearable technology and gesture computing opens up the gate for ultra-modern applications that goes beyond regular applications on both sides. This combination especially gives a direct control over human–machine interface in any form, omitting away all the conventional forms of communications carried out earlier.

As discussed earlier in this book, the development of mobile phones with apps representing many add-on devices is the close parallel technology to wearable computing. This is made possible by the use of incorporation of different sensors in a one simple handheld device. If mobile phone sensors have any applications to measure human activity and human activity recognition (HAR) is currently an active research area using the sensors included in the

DOI: 10.1201/9781003052906-10

phones. Although there are few applications available on simple human-activity recognition like walking, running, jumping, etc., a little research has been conducted on complex human activity recognition.

Conventional machine learning approaches failed in recognizing human activities performed as different actions in a day since most of them are complex actions. It is a common practice to deploy multiple sensors for the recognition of complex activities since it is very difficult to understand complex human activities with high accuracy using a single sensor. A research was conducted using multilayer LSTM on different publicly available dataset such as, PAMAP2 and WISDM that contain complex human Activities. The LSTM-based approach showed that it can achieve good results using single sensor data instead of multiple sensors, but at the same time it also proved that using multiple sensors increased the accuracy of each activity. This approach achieved the high accuracy of results when compared with CNN and other traditional machine learning approaches.

10.2 Magic Ring for Recognition

A self-sticking paper band on the wrist is commonly used as an authentication to entries of games and fun activities in amusement parts. A much better version of the band is the plastic wrist band or a down-sized magic ring that goes beyond mere authentication to switch on-off circuits, etc. It is possible to make the ring to communicate with devices with a passcode not automatically but with the consent and efforts of the wearer to make it safer so that anyone who steals or borrows the ring may not use it as its owner.

10.3 Authentication of Sex Robot Access Using Wearables

Wearables are freely available everywhere as consumables and counter sale products. (Figure 10.1). It is necessary to make sure that no pirated version of hardware lock, or crack software, etc. are available for juveniles in future as purchasable items to misuse them to gain access to sex robots. But it is very difficult to ensure that and history had shown the failure of such security mechanisms. Hence, it is highly recommended to have biometric protection which comes in protection of juveniles from unnatural affection with the Humanoid Robots of the Future. The WT enhances that biometric locks to make it difficult for unauthorized usage or intentional lending misuse.

FIGURE 10.1
Shops selling assorted wearables in open market

10.3.1 The Problem of the Future Generation

Everyday news about unlimited humanoid personal robots and how they are flooding into the market by many companies are reported. In the beginning, the purposes assigned for such robotic torrential are like helping humans in house or office jobs. But later they extend their services by supporting humans in the health treatments, emotional pacifications, daily life, special operations for disabled, unconventional works, as well as for having sex. This conception seems to be bringing new advantages, fun, thrill, and freedom, but there exists dangers like total dependency, addiction, sociological disorders, and misuse. There are possibilities to eradicate these adverse efforts, and we need novel technologies to provide a trivial contribution in curbing the illicit usage of assistive and personal robots for dangerous activities of crime, including having inappropriate sexual relations like rape and pedophilia. The security measures need to be taken with a paradigm shift in technology directions such as DEEP learning and not just granting authentication to control robots with biometrics and conventional passwords.

The fast growing wearable technology contributes in preparing the human race to accept the reality that they are going to live with machines which are all, readily available, and becoming stronger, faster, and highly talented, and even, more beautiful. We learn from news that human-versus-machine programmes involves AI systems that is argumentative by which many live debates were conducted. Those logical arguments under titles like 'space exploration should be subsidized', show the efficient performance of AI when an opponent human disagrees, the AI program suitably offered a rebuttal. Recently this debate, which went well between an IBM computer AI program called 'Project Debater' and several human participants, is another proof that artificially intelligent machines are making progress at skills, such as arguing, which previously were reserved for people.

But now technology has to be developed beyond those superiorities already exhibited by the robots and to focus on its questionable character, the 'high emotional attachment' which the robots can create in their human counterparts. The aim of this technology in demand is to develop an inbuilt restraint mechanism to be included in the brain activities of the robot design which is above a multimodal biometric authentication to have a natural solution similar to the ethical behaviors of human families and society.

The emotion and emotional attachment start from the human side on robots when they become closer to humans in the form of personal robots. It is human nature to develop a sense called belongingness with every object they deal with, just as children develop a bond with their toys. Research on assistive robots, robo-gender, sexbots, robo-satisfaction, neuro-dildos, is already in place and directly leading with the affective components between humans and robots.

The robotic dependency arrives in different forms. The basic robotic dependency will arrive on the continuity of getting the robotic assistance by the disabled and elderly people. The next is when a common person requires wearable robots for special occasions or when a strategic commando seeks robotic help during special operations like rescue etc. In due course, this will become a habit and as simple as the calculator dependency among the school students for their arithmetic calculations nowadays.

When there is a dependency, there generates a bond of belongingness and possessiveness in due course. Obviously people have the habit of maintenance of their belongings. It includes basic cleaning; proper technical maintenance, and it goes up to decoration. According to Susan Spicer, an expert, children getting affectionate to nonliving things, such as blankets and dolls, is natural. She found it as common and perfectly healthy for toddlers to develop a deep attachment toward 'transitional object' as called by the psychologists. This childhood habit continues to the adulthood and develops a stronger bond between the close things that help, assist, protect, and share life with them. More on this, says, Stanley Greenspan a medical expert, that the babies develop affection when they attain as early as 2–4 months of age.

The same attitude can be seen among grown up boys who love motor cycles, and the men on cars. The emotional attachment comes for men when a single object can bring them both, the social recognition and satisfactory instincts.

It is felt nowadays there are strong possibilities of making an emotional bond with the robos that is similar to the bonds children make with their teddies and lively toys. This article is to create awareness and to lead further research and security precautions in the forthcoming robotic investigations.

The concern is to fight the most dangerous part, the unauthenticated and misuse of robots for criminal purposes as well as purposes against the law and order. So analysis and precautions are suggested with the context of personal activities.

10.3.2 The Reason for Focus Toward Robotic Emotions

Scientists predict that the rise of artificial intelligence (AI) will influence the speedup of researches like personal robotics, which add human-like qualities to sex robots that make them seem more 'factual'. The inclusion of virtual and/or augmented reality technologies makes the possibilities of realistic sex robots most attractive and realistic which places human life under disaster.

Laws in many countries are regulating the sexual dependency in a systematic way from ages past. In India, Indian Penal Code 377 criminalizes such sexual activities as against the order of nature. Even the sale, advertisement, distribution, and the public exhibition of obscene books, sketch, drawing, or any other 'obscene' object, is made illegal by Section 292 of Indian Penal Code (IPC). People are asking that the obscene law has to be fortified and usage of such toys must be curbed. Selling such devices and robots online as well as in shops needs to be banned to save the humans from a deadly character disaster. Raising concern from parents and general public leads the researchers and authorities to pay attention and focus in this area. It is being felt mandatory to form fortified regulations at the earliest.

10.3.3 Related Examples

Experiments by Nadia Magnanet Thalmann of Miralab, using her Cloyi, and Cynthia Breazeal of MIT artificial Intelligence lab, using her face robot called Kismet, were conducted with human subject interaction. Mark1, Bina48, Jia Jia by Xiaoping are additional examples for related robots.

Sophia a full humanoid robo who was activated in year 2015 was granted a questionable citizenship by Saudi Arabia in year 2017. Sophia was manufactured at Hanson Robotics in 2015 to be media-savvy with exceptional conversational skills. She can show around 50–62 facial expressions by the actuators embedded in the face of the robot. It is to naturalize the robot behavior to that of humans since girls are expected to show more empathy and sympathy than boys both in the form of facial emotional displays and empathic behaviors. Its producer Hanson says that the realistic design of Sophia is intended to allow robots to form meaningful associations with the humans.

Many questions are unanswered on Sophia and the business and publicity motived citizenship granted on her. Is there any difference in the granted rights on Sophia or has she been given the same rights as other human citizens? Would it be illegal for her to go out without mask and what if someone keeps her confined and forces her to do typical robotic tasks like cleaning house or building products at factory? Is she having voting rights? Is she allowed to get married on her free will? Is she allowed to approach the court for divorce? If divorced by spouse, what will be the compensations? Would it be a crime of murder to turn her off or erase her memory? And what if someone tried to do that and she resisted? Would she be allowed to fight for survival even if it means going on the run, killing her maker.

10.3.4 The Era of Personal Robots and Their Characters

10.3.4.1 Need for Personal Robots

Personal robots are necessary in future to support as people grow in age, even for youngsters who undertake adventurous and risky operations alone. It is unavoidable for disabled people to make use of a personal robot which can understand their requirement according to their movements and affective expressions etc. (e.g. Secret, protection).

The Assistive robots reduce the tedious workload of caretakers and do an amazing work by even cleaning the humans and their wounds when necessary without hesitation and aversion. It is estimated in Japan, by 2025, more than 80% of elderly care would be done by robots, not by the caregivers.

Wearable robots, for the purpose of enhancing the human capability, are making humans a hybrid cyborg and elevate them out of the limitations. A rescue robot is essentially needed and is a welcomed invention to help humans in multiple dangers. A wearable robot can enrich the camouflaging capacity of humans since the silicone skin can take the form of its surroundings and terrain.

Once a competition for achieving predefined goals is set, the robots always try to outperform humans by the accumulation and enrichment of components present in them. In social interaction, the parameters like beauty, sexual attractiveness, social attractiveness, vivaciousness and energy, presentation, and sexuality are now the goals of robots to push back humans. Everyday proximity to social robots is inevitable in the near future, and appreciation of their services will develop into them a sense of being part of human teams. Since robotic technology and artificial intelligence are moving at a great speed, in the future, one's robot teammate will possibly be the friend that one shares secrets and dreams with. The robots in near future will attain the evolution of identifying their human friends by simply touching a part of their body, or smelling them, as a scenario, this might sound unimaginable in reality, but, in the next decade, humans will encounter robotic personalities that are remarkably similar to those of humans, and these robots will certainly be highly competent. With time, these entities will not only answer us accurately, logically, and knowledgeably, but also laugh at our humor and reciprocate our feelings. The robots will develop into personas that one confides in, become partners with, falls in love with, wants to be intimate with, and bonds with in a legal union.

10.3.4.2 Gender Approaches – Transgender Robots

When it comes to the gender, both male and female personal robots are available to suit the human partner. Men have a greater interest in visual sexual stimuli than women. Attraction between genders comes from visual, audition, and olfaction criteria. Many men and women have different feel in selecting their partners at the first sight. There is also transgender robot option available from the manufacturers for people who have the tendencies to being attracted to transgender (Figure 10.2).

The differences, as well as alternate combination in male/female dichotomy, are liked by them and they are interested in explicit focus on the transgender body and appearance. Tactile attraction is above all of them since the tangible shape and size matter more.

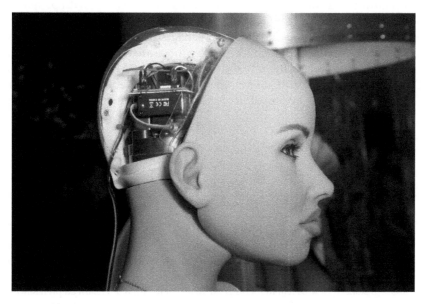

FIGURE 10.2
Different faces of Harmony created by Matt McMullen in different looks and hairstyles

10.3.5 Is It Possible to Have Affective Approaches with Robots?

10.3.5.1 Creation of Feelings

Apart from Sophia, many other humanoid robots like Alice, Albert Einstein, Jules, zeno, cylon etc. started coming into the human world to interact with us and the coexistence has already started whether we, humans, like it or not. The interactions include expression of affection between each other category. Robotic hug is considered as a way of treatment and is to support humans when they are emotionally down. Robots called Thomas and Janet were taught how to kiss. But there's never a stage at which the robots know what they are talking about; whereas humans are overpowered by their feelings, impressed by the performance etc.

Even reports about Sophia, when it received a kiss in public as an appreciation by a lady for the high relevant talks she performed, are available. This is a great example how humans easily get attached to robots even after knowing it as a nonliving thing. Affection shown in hugs, kisses etc. is generated as an appreciation, admiration, and belongingness. Even though the robots like Sofia don't feel the affection, they started to respond in a receptive way and make the tempo alive (Table 10.1).

Sofia responded in words to the affection shown and in later days, we can bring in such responses in movements, reciprocal hugs, kisses, and other public and private responses.

Many sex robots like Samantha from Synthea and Harmony from Realbotix with beautiful face and organs started coming into existence with people and accompany them in outings too. Many such videos are already available in the media and these bots have special sensors on their private parts. They are programmed to react when touched in those areas and mimic the feeling of being with human partners to the user. Harmony AI works with Android 5.0+ OS, Version 1.3 with a memory size 250MB presently and can improve to higher levels. They are capable of measuring human arousals by reading

TABLE 10.1

Various levels of sex equipment and their functions

S. No.	Equipment	Functionality	Salient features
1	Sex toys	Replicas of human, passive and needs humans to use it.	Texture, materials, and shapes
2	Mechanized toys	Along with conventional toys, motorized and freedom of usage.	Speed variants and chargeable
3.	Intelligent toys	Along with mechanized advantages, learn to adopt the customization.	Brings more suitability and convenience
4	Sexbots	A real human size, with all color variants of features, more intelligence infused in identifying partner.	Learn all preferences and desires, Higher level cooperation.

the body parameters by means of sensors including embedded penile plethysmography and react according to it. Even though remote controlled or programmable virtual sex for men through wearables is possible but more toward usual archetype, this intelligent electronic sex hardware starts to learn the habits, sequence, and appropriate responses from the human partners and increase their erotic capital. They make themselves suitable to desired movements and rhythms of human partners and are available at their service all the 24 X 7 hours without any fuss and create an addiction toward them. Samantha may be made exception to say 'No', but there are alternate options like having one more Samantha or some other robo for instant remedy or to modify Samantha's network.

10.3.5.2 Appreciation and Rewards: How Robots View This?

Using deep learning principles, the appreciations can be converted into weights from which the future decisions of the robots get influenced. So indirectly the appreciations are having impact on robotic brain as they have in humans. In due course, the robots identify human appreciation looks towards them, and readily adjust themselves to human mood. Artificial seductive smelling perfumes like pheromone which has variants with compounds like Androstadienone, Androstenol, Androstenone, and Androsterone will be included in the robotic character along with secreting lubricants. This further makes the personal robots more attractive and seductive. Experts say that this may lead the robots to display novelty, excitement, and variety in due course. Robots also can be converted to show interest in sex by such appreciations which comes from the basis of Reinforcement Learning.

10.3.5.3 Psychological and Sociological Approaches

Sex robotic addiction is seen as a higher level of deviation from the natural attraction and affection between humans. It may not be considered as alternative lifestyle. Many studies show that these personal robos aggravate loneliness rather than solving it.

Family cannot be formed with robots as in the case of marriage with a transgender. Even if a child is adopted as in the case of families with transgender partners, a child won't accept a robo as its parent. This will cause a severe psychological problem in the children and their future. It is dangerous for children to misuse or get misused by these robots and children are highly vulnerable to this avatars.

Already, several studies suggest that gay men, lesbians, and bisexuals appear to have higher rates of some mental disorders compared with heterosexuals. Until now the number of cases for suicide attempts is higher in the case of gays and lesbians when compared to heterosexuals. So a serious outbreak of psychological and sociological disorders is expected when the robos start sharing bed with the humans. It is questionable and creates lot of confusion and ethical conflicts if a person creates a personal robot which looks like neighbors' wife or a famous politician or a celebrity.

Malaysia banned the second international congress on 'Love and sex with robos' in 2016 to avoid controversies and explicitly showing its stand against such developments in the society.

10.3.5.4 Rights of Robots? and Robo-ethics

The power of the robots is their accumulated knowledge bank gained through the days of their activation by being in operation of their domain. The treat to the robots is for being stripped off of their memory purposely or accidentally. Any unauthorized removal of their growing brain, or accidental flashing need to be avoided to protect the robots. When the brain is flashed the robot become inefficient and outdated and causes the death of the robots from being active in their domain.

When the robots comes into coexist with humans they also have the right to share our constitution to a certain level as right of protection, right against misuse, right against deterioration, rights against damage, and unauthorized use.

When the rights are protected, it is easy to control the order of coexistence and avoid unwanted crimes against robos or crimes misusing robos against humans or other robos. A deep concern is building up that the robos should not identify themselves as a member of a particular caste or religion. More worries buildup while thinking about such next level transformations of social robots carrying pistols and pepper spray as the military robots doing in Israel.

10.3.6 Areas Where Humans Excel the Robots

10.3.6.1 Preach – Cannot Repent or Realize

Any kind of message either technical or spiritual can be delivered interestingly for hours by the robots with illustrations and heart-touching narrations. Robots these days are given a Scottish accent to talk, since makers of such robots feel it is less robotic and more humanized. The impact of hearing these talks is only on the human side. There won't be any decision making or change in behavior or attitude in the robots if they hear a preaching of human or other robots.

10.3.6.2 Indulge Carnal Pleasure – Cannot Produce Own Children

Hypothetical artificial wombs may be fitted in the personal robots in future. They can do the work of an incubator for the neonatal or for a fetus. But there won't be a chance of a robot producing an embryo of its own, since there is no life or DNA for robots. They can only survive in the level of sex machines and personal cohorts [6]. In no way robots can replace a mother or faithful wives or husbands. So, an alternative that leads to psychological disorders can never be a replacement.

10.3.6.3 Remember – Cannot Imagine

Researchers like Dr. Aude Oliva are working on this synthetic imagination called machine imagination. The machine imagination uses tools and insights from interdisciplinary fields like Computing, Artificial Intelligence, Rhetoric, Psychology, Neuroscience, Creative arts, Philosophy, Affective computing, Cognitive science, Linguistics, Operations research, Creative writing, Probability and Logic. But so far, no framework or critical findings exist and it is impossible to achieve the imagination of man in every dimension and direction. If given a paint brush and a caption and coverage, the robots can't draw an art covering the caption by its own.

10.3.6.4 Say a Lot of Stories – Cannot Write One by Own

When there is no imagination, own stories never arise. Mimicking natural stories can happen with ennui repetitions but it won't be a perfect story of novelty and timeliness. A robot cannot be perfect in making a crying child to sleep by telling a bedtime story. The process incorporates so many interferences from the child by means of questions about the narrations and physical fracas.

10.3.6.5 Sense and Show Expressions – Cannot Feel Anything, Empathy Sympathy Joy, etc.

Sophia, a life size humanoid robot activated in year 2015, was manufactured at Hanson Robotics in 2015, with a focus of making it a media-savvy with exceptional conversational skills. She can show around 50–62 facial expressions by the actuators embedded in the face of the robot. It is to naturalize the robot behavior to that of humans, since girls are expected to show more expressions than boys in facial display. Its producer, Hanson, says that the realistic design of Sophia is intended to allow robots to form meaningful associations with the humans.

The number of possible facial expressions neither introduces feelings nor increases understanding of feelings. They cannot understand your mood if you change your facial expressions differently and take away behind your tone and face.

For example, a human can identify and understand while a person is talking over phone without the call connected the other side, which robots cannot detect.

10.3.6.6 Restricted to Only the Language of Operation – Humans Can Learn an Unknown Language with Time

Language is a gift of God to mankind and humans can understand more than one language automatically if they are exposed to hear such dialects. Whereas this natural habit of learning coordination based on events and situations is limited in the case of robots.

10.3.6.7 Can Control Animals – Cannot Pet or Dominate Animals

Humans control animals in close proximity by their whole presence as well as in remote places by the fitted wireless sets on them. The affection and control shown by the humans on the animals can be exercised by the robots, but is limited. The animals smell the humans to identify them. The smell of persons is unique and different from each other. Even same perfume smells different on different persons. [5] Though robots may smell silicone, it will not be unique to a single robot. The robots cannot show real affection to the animals and can't relate the animal relationships. Moreover, humans know how to react in the wild with the wild animals and humans possess inherent survival technology. Even the robots know how to exhibit taekwondo movements; they cannot protect themselves from angry wild animals and natural disasters.

10.3.6.8 Can Do the Work Given – Can't Act in Different Roles and Not Trustable

There are no different roles robots play with different people and they lack discerning knowledge between genders, elders, and youngsters. The robots are not equipped with parenting skills, though they can take care of children. The information given to robots can be taken by any intruder and it can be retrained to forget what it was trained. So, it is worthless leaving any confidential information with the robots as robots lack the ability to retain them as humans do.

10.3.6.9 Can Give Lecture – Cannot Teach Various Children According to Their Knowledge and Intelligence Capacity

As Sophia relates with social encounters and interacts with everyone relevantly, the personal robots can identify the face, touch, and smell of people who interact with them and can remember much information about the persons who it comes across. When it comes to topics, it's easy for the robots to act as an expert system and cross refer the Wikipedia based on the questions posed to it. They can give an illusion that they know everything in the subject matter. But when it comes to teaching, the robots lack personalization and customization in teaching students of different caliber and understanding their intentions when they ask irrelevant questions. This deeper understanding of each of the students and their capacity and the reasoning that varies in students is not possible for a robot. In addition, robots cannot distinguish between a sarcastic statement and a normal statement in general topics. Though researches are in progress to identify sarcasms in particular domains, humans can identify it universally.

10.3.6.10 Can Travel – Cannot Do a Solo Travel to an Unplanned Location and Come Back

Solo travel refers to a travel to a place alone (or with pets) where one will spend a significant duration of time after getting there with no one. It is not just driving or flying from point A to point B, to meet family or attending an event, but this is to earn more thrilling and adventurous experiences, as well as learning new skills of survival and task accomplishments. Many people during such trips earn friends, knowledge, and even money. The robots can never undertake such unplanned trips and never accomplish an undefined task.

10.3.7 Solution and Research Direction

An authorization component has to be inbuilt with the robots using the deep learning concept which helps the robots to 'learn like' humans. Robots cannot be misused as a dump, slave devices, especially in the case of personal robots. The module representing the authorization for access or activity should be capable of discerning the rationale behind the command given to it. This will help the personal robots to be used for the right purpose by the right people.

This is a deep learning way of making the robots to identify authorized persons as humans know each other. Password, pin number, or biometric-based authentication is a dump machine approach, whereas the proposed work exhibits the deep learning approach. In this way robots identify the person and relate them as humans do. The machine learning takes place from the touch, speech, context, approach, smell, and activities. Others can never get access to those robotic services even if desired so. The possibilities are studied and are found feasible for the robots to identify their disabled partners using deep learning which is more accurate and robust than simple biometric authentications. Positive uses of personal robots are mentioned in the conclusion.

The solution presented in this work is a simple idea to prevent the seamless misusage of personal robots by providing a check measure in the brain of the robot through modifications in AI program for eliminating the unauthorized people or activity in the initial stages. People claim that personal robots may be used for birth control and reduction in sexually transmitted diseases with proper cleaning mechanism. Some may argue that is the only remedy for people with permanent disabilities who find it difficult to marry a human partner. In general, it is unethical to keep a sexrobo in a family for common people and the institution of marriage and family should not be placed under threat by such robots. A severe psychological damage may happen to users, and recovery is extraordinarily difficult. These inventions with innovations influence the perverse minds and could forever change the nature of human existence. The basic and important things of a marriage are to produce children, but before that there must be a conscious consent to get married to each other. That is the reason the age for marriage is regulated so that both the partners can consent to the marriage on their free will. Even if an artificial womb is created to produce babies, it will not create the baby of the robo–human fusion. Secondly, robots can never have their own free will to give consent. The increase of perverted feminism, mental sickness, damage to sexuality, and damage to sex organs and health will be resulted in human society that adopts the digi-sexuality. Even people argue that child sexrobos can prevent pedophilia, but it is not a justifiable argument and can cause reverse effect on humans to move from robots to children and for promoting pedophilia. The death of a robot makes adversary results in humans when the robot starts misbehaving from the routine after the death reboot. Lawmakers and corporations in the near future have to enforce legal and ethical suppression of machine emotional maturity so that people can feel safe.

There are positive ways of personal robots' use in anti drug addiction, recovery from fear, assistive help including reading robots, wheel chair, and walking assist. It is so good and tasty to use the chef robot developed by Moley Robotics who can cook 2000 varieties of food at present.

With much research and secure implementation, robots can be used in military where the advantage of loss of life and information retrieval by tortures etc. are neutralized, which are the threats a human soldier undergoes.

11

Soft Cyborgs and Cyber Physical Systems by WT

The overview of the human machine service application with a biosignal-based cyborg activity is to enable all clauses of human beings to elevate their living conditions and life style. Any compromise or a poor design in human–machine interfaces can lead to many unexpected problems. In real world a classic example of this is the Three Mile Island accident happened in Pennsylvania, a nuclear meltdown, where investigations concluded that the design of the human–machine interface was at least partially responsible for the disaster. The same way, accidents in aviation have resulted from manufacturers' decisions to use non-standard flight instrument or throttle quadrant layouts: even though the new designs were proposed to be superior in regard to basic human–machine interaction, pilots had already ingrained the 'standard' layout and thus the conceptually good idea actually had undesirable results. Wearable technology plays its role in a well-advanced alternative to the conventional cyborgs and cyber physical systems.

11.1 Soft Robotics

The soft robotics is a new generation branch of robotics where robots are created with a different material to mimic the characteristics of any living organism. This bio-inspired creativity is not only for the aesthetic change of a steel look nature of robots but to make the robots more appropriate to the application and its environment. The advantages of soft robotics are ranging from enhanced grasping to 3D perception, high performance, improved pick-place tasks on delicate and sporadic products.

Soft robotics is an interscience child which involves computer science, chemistry, plastic and rubber technology, and mechatronics. Fluidic actuators are used to get fit into expanding bladder-like robotic elements. Even, vacuum, collapse, and bulging mechanisms are incorporated into the functions of these flexible, soft structures. Catalyzed polymers like silicone rubber are widely used in this new generation soft robotic structures to bring them as close as to the biological look. Further to its improvements toward durability, fiber composite materials are in use.

Mainly, it is necessary for the humans, for not get hurt or fall by the robust robots while they move along the humans in a busy work place. Moreover, soft robots have the flexibility better than the humans and transforming robots to move into many types of obstructions which humans cannot do during a rescue operation to save a life.

11.2 Cyborgs

Wearable computing talks about the novel initiatives like 'invisible man' of Professor Susumu Tachi and cyborg of Professor Kevin Warwick. These researches may bring a shock of alienism to an amateur researcher since they may doubt its immediate potential for commercial and practical breakthrough. As many are aware of the cyborg, it's a cybernetic organism, a typical example of a wearable, where machines are infused into human body to give man the supernatural powers and the machine a super intelligence. Already cyborgs started walking in the minds and cartoons of Indian children; the real-life usage of cyborg concept is limited to pace makers and artificial limbs as on date. Scientists and their professors who involved in the academic wearable computing projects often end up in newspaper columns for publicity reasons rather than in a complete end product.

The ever-growing processing power according to Moore's law poses a threat to human race on the shadow. The machines become stronger and stronger, more accurate, and faster. The control systems for such machines are also in a stage of evolution to achieve the precision of control. The reliability of any external system is not perfect on temporal or spatial aspects. The ergonomic support is achieved with the development of wearable computing. With the help of wearable computing, we can explore the possibilities of giving a paradigm shift to the cyborg concept and make the participation of such cyborgs on our future social life with all their benefits to humans. The alternative technical approach toward noninvasive cyborgs will also be discussed in the coming sessions.

A remarkable breakthrough came with the initiatives of Professor Kevin Warwick when he implanted a silicon chip transponder with a size of 25 mm (equal to the size of a British 2 pence coin) in his forearm as a part of his experiments in 1998 with many permissions from the authorities. He produced a 64 bit data from the transponder to be received for processing by the computers. Later, in 2002, a one hundred electrode array was surgically implanted into the median nerve fibers of the left arm of Professor Kevin Warwick to act as an interface with his nervous system and a computer.

As an outcome of such experiments, most notably Professor Warwick controlled an electric wheelchair and an intelligent artificial hand, developed by Dr. Peter Kyberd, using this neural interface.

In addition to measuring the nerve signals transmitted along the nerve fibers, the implant was also able to create artificial sensation. This bi-directional functionality was demonstrated with the aid of Professor Kevin's wife Irena and a second, less complex implant connecting to her nervous system.

Such experiments created lot of enthusiasm and interest in researchers and affective computing started different dimensions after this incident. Ambient intelligence and ubiquitous computing are the next areas that started contributing to this development further. Motivated by such exhilaration, I decided to make myself also into a cyborg but without taking those painful efforts of Professor Kevin Warwick. Hence, I developed a concept called soft cyborg with the help of wearable technology.

11.2.1 Hard Cyborg and Soft Cyborg

Is a hard cyborg being really hard like a metallic robot? And a soft cyborg is soft as a soft fur doll or jelly filled balloons that we see in soft robotics? No, it's just a technical classification like hard wired circuits vs. portability of software (Figure 11.1).

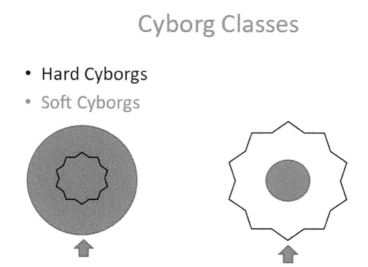

Cyborg Classes

- Hard Cyborgs
- Soft Cyborgs

Machine inside man (Hard) Man Inside Machine (Soft)

FIGURE 11.1
Classes of Cyborg

The difference between Soft Robotics and soft cyborgs are

- Not of same concept
- Not of same Mechanism
- Not of same Life and Machine combination
- Not of Convertible nature

The commonalities of Soft Robotics and soft cyborgs are that they both use

- Soft stretch sensors, bending sensors, pressure sensors, force sensors
- Both use similar interfacing style
- Many utility components can be embedded on both
- Soft robotics can also follow the wearable technology like soft cyborgs.

Professor Kevin extended his experiments beyond the cyborgs to create an artificial brain or an independent brain of a person out of the body for remote sensing into the primary brain. While I moved on to the soft cyborgs, with the 'Spiderman', 'Superman', concepts in which a common man's ability and power are augmented when they wear their magic suits.

While it was interesting, to do a series of applications with the help of wearable technology, under the soft cyborg title, I found a threat to the safety and security of the user, since my primary aim is the dual application of every device as an assistive technology and augmentative technology (cyborg Technology) which goes into authentications as a part of it. So it was decided to make the wearable devices into smart wearables which can only operate on its owners and no one else can operate them even if they wear it. (A video of such example can be found from https://www.youtube.com/watch?v=Gy0WDTki7eA). Biometric recognition and recognition of owner using technologies beyond biometrics are found necessary when people intend to use personal robots which are already discussed in this book (Figure 11.2).

11.3 Cyber Physical Systems

Cyber physical systems are working on a more comprehensive approach than embedded systems, real-time systems, communications, or control. In spirit, it is close to mechatronics in control since it is being interdisciplinary, but CPS is intended to be much more interdisciplinary.

It appears that IoT puts more focus on the first steps of the adaptation cycle: sensing, monitoring, and data analytics. CPS focuses more on the later part

FIGURE 11.2
Two Professors as an example of Soft and Hard cyborgs Known by their Experiments - The Author with Prof.Kevin warwick.

of the cycle: reasoning, planning, and actuation. The definition of CPS gives a picture that smart systems with sophisticated software sit on the physical systems, and this certainly resembles the soft cyborgs and hence is worth exploring into CPS for enhancing WT and assistive technology.

The difference between an embedded system and CPS is that an embedded system is a computer system with a dedicated function within a larger mechanical or electrical system, often with real-time computing constraints. It is *embedded* as part of a complete device often including hardware and mechanical parts. It's not having IoT options (e.g.EVM).

A CPS is a complete system by itself and is always real time, pervasive, autonomous, and always includes hardware and mechanical parts. Computers and physical systems are tightly coupled, and timing is critical. CPS works in physical environment.

A CPS is an all-in-one system that has no problems in the format of interfacing. It includes all, digital, analog, physical, and other possible human interfaces like the Aura that we have seen in the mechanical mirror system can understand each other and work without any difficulty and claim the name smart for all its component systems.